Collins

GCSE 9-1
Biology
in a week

Tom Adams

Revision Planner

DAY 1

Page	Time	Topic	Date	Time	Completed
4	30 minutes	Structure of cells			
6	35 minutes	Organisation and differentiation			
8	40 minutes	Microscopy and microorganisms			
10	40 minutes	Cell division			
12	45 minutes	Metabolism – respiration			
14	40 minutes	Metabolism – enzymes			

DAY 2

Page	Time	Topic	Date	Time	Completed
16	35 minutes	Cell transport			
18	25 minutes	Plant tissues, organs and systems			
20	40 minutes	Transport in plants			
22	40 minutes	Transport in humans 1			
24	40 minutes	Transport in humans 2			
26	40 minutes	Photosynthesis			

DAY 3

Page	Time	Topic	Date	Time	Completed
28	45 minutes	Non-communicable diseases			
30	30 minutes	Communicable diseases			
32	40 minutes	Human defences			
34	40 minutes	Fighting disease			
36	45 minutes	Discovery and development of drugs			
38	30 minutes	Plant diseases			

Page	Time	Topic	Date	Time	Completed
40	25 minutes	Homeostasis and negative feedback			
42	35 minutes	The nervous system			
44	40 minutes	Nervous control			
46	45 minutes	The eye			
48	35 minutes	The endocrine system			
50	45 minutes	Water and nitrogen balance			

Note: You will see the following logos throughout this book:

HT – indicates content that is Higher Tier only.

WS – indicates 'Working Scientifically' content, which covers practical skills and data-related concepts.

Structure of cells

Prokaryotes

Prokaryotes are simple cells such as bacteria and archaea. Archaea are ancient bacteria with different types of cell wall and genetic code to other bacteria. They include microorganisms that can use methane or sulfur as a means of nutrition (methanogens and thermoacidophiles).

Prokaryotic cells:

- are smaller than eukaryotic cells
- have cytoplasm, a cell membrane and a cell wall
- have genetic material in the form of a DNA loop, together with rings of DNA called plasmids
- do not have a nucleus.

The chart below shows the relative sizes of prokaryotic and eukaryotic cells, together with the magnification range of different viewing devices.

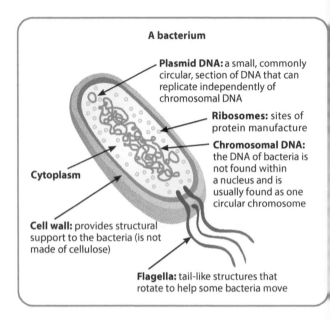

A bacterium

Plasmid DNA: a small, commonly circular, section of DNA that can replicate independently of chromosomal DNA

Ribosomes: sites of protein manufacture

Chromosomal DNA: the DNA of bacteria is not found within a nucleus and is usually found as one circular chromosome

Cytoplasm

Cell wall: provides structural support to the bacteria (is not made of cellulose)

Flagella: tail-like structures that rotate to help some bacteria move

| 1 nm | 10 nm | 100 nm | 1 µm | 10 µm | 100 µm | 1 mm |

Prokaryotic cells

Eukaryotic cells

Key
nm = nanometres
µm = micrometres

Eye — Visible with

Light microscope — Visible with

Electron microscope — Visible with

mm → µm
×1000

µm → nm
×1000

Human cheek cell (typical animal cell)

Palisade cell (typical plant cell)

Mitochondrion: where aerobic respiration takes place

Cell membrane: controls movement into and out of the cell; contains receptor molecules and is a selective barrier

Cytoplasm: chemical reactions take place here

Nucleus: controls what the cell does and contains the genetic information

Ribosome: where protein synthesis occurs

Cellulose cell wall: strengthens and supports the cell

Chloroplast: contains chlorophyll and absorbs light for photosynthesis

Permanent vacuole: filled with cell sap, it helps to support the cell

Eukaryotes

Eukaryotic cells:

- are more complex than prokaryotic cells (they have a cell membrane, cytoplasm and genetic material enclosed in a nucleus)
- are found in animals, plants, fungi (e.g. toadstools, yeasts, moulds) and protists (e.g. amoeba)
- contain membrane-bound structures called **organelles**, where specific functions are carried out.

Pictured above are the main organelles in plant and animal cells.

Plant cells tend to be more regular in shape than animal cells. They have additional structures: cell wall, sap vacuole and sometimes chloroplasts.

QUESTIONS

QUICK TEST

1. Prokaryotic cells are more complex than eukaryotic cells. True or false?

2. All cells have a nucleus. True or false?

3. Which organelle carries out the function of protein synthesis?

EXAM PRACTICE

1. a) Bacteria contain circular sections of DNA within their cells.
   Name these structures. **[1 mark]**

   b) Plant cells differ to animal cells in which one of the following ways? Tick **one** box. **[1 mark]**

   They contain a nucleus. ☐
   They do not have a cell membrane. ☐
   They possess a cell wall. ☐
   They contain ribosomes. ☐

   c) Muscle cells contain many mitochondria.
   Explain why this is. **[2 marks]**

   d) The average magnified length of a bacterium is 10 mm.

   Using a magnification of 500, calculate its **actual** length in **µm**
   (1 mm = 1000 micrometres)
   Show your working. **[2 marks]**

SUMMARY

- **Cells are the basis of life. All processes of life take place within them.**

- **There are many different kinds of specialised cell but they all have several common features.**

- **The two main categories of cells are prokaryotic (prokaryotes) and eukaryotic (eukaryotes).**

Organisation and differentiation

Cell specialisation

Animals and plants have many different types of cells. Each cell is adapted to carry out a specific function. Some cells can act independently, e.g. white blood cells, but most operate together as tissues.

Type of specialised animal cell

Sperm cells ● They are adapted for swimming in the female reproductive system – mitochondria in the neck release energy for swimming. ● They are adapted for carrying out fertilisation with an egg cell – the acrosome contains enzymes for digestion of the ovum's outer protective cells at fertilisation.	
Egg cells (ova) ● They are very large in order to carry food reserves for the developing embryo. ● After fertilisation, the cell membrane changes and locks out other sperm.	
Ciliated epithelial cells ● They line the respiratory passages and help protect the lungs against dust and microorganisms.	
Nerve cells ● They have long, slender extensions called axons that carry nerve impulses.	
Muscle cells ● They are able to contract (shorten) to bring about the movement of limbs.	

Type of specialised plant cell

Root hair cells ● They have tiny, hair-like extensions. These increase the surface area of roots to help with the absorption of water and minerals.	
Xylem ● They are long, thin, hollow cells. Their shape helps with the transport of water through the stem, roots and leaves.	
Phloem ● They are long, thin cells with pores in the end walls. Their structure helps the cell sap move from one phloem cell to the next.	

Principles of organisation

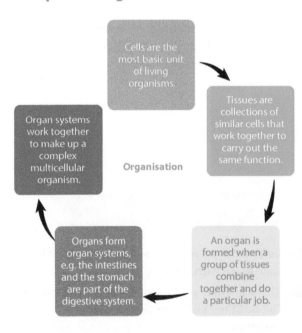

Cells are the most basic unit of living organisms.

Tissues are collections of similar cells that work together to carry out the same function.

An organ is formed when a group of tissues combine together and do a particular job.

Organs form organ systems, e.g. the intestines and the stomach are part of the digestive system.

Organ systems work together to make up a complex multicellular organism.

Organisation

Cell differentiation and stem cells

Stem cells are found in animals and plants. They are unspecialised or **undifferentiated**, which means that they have the potential to become almost any kind of cell. Once the cell is fully specialised, it will possess sub-cellular organelles specific to the function of that cell. Embryonic stem cells, compared with those found in adults' bone marrow, are more flexible in terms of what they can become.

Animal cells are mainly restricted to repair and replacement in later life. Plants retain their ability to **differentiate** (specialise) throughout life.

Using stem cells in humans

Therapeutic cloning treats conditions such as diabetes. Healthy pancreas cells can be cloned in order to achieve this. Embryonic stem cells (that can specialise into any type of cell) are produced with the same genes as the patient. If these are introduced into the body, they are not usually rejected.

Treating paralysis is possible using stem cells that are capable of differentiating into new nerve cells. The brain uses these new cells to transmit nervous impulses to the patient's muscles.

There are benefits and objections to using stem cells.

Benefits	Risks and objections
Stem cells left over from in vitro fertilisation (IVF) treatment (that would otherwise be destroyed) can be used to treat serious conditions.	Some people believe that an embryo at any age is a human being and so should not be used to grow cells or be experimented on.
Stem cells are useful in studying how cell division goes wrong, e.g. cancer.	One potential risk of using stem cells is transferring viral infections.
In the future, stem cells could be used to grow new organs for transplantation.	If stem cells are used in an operation, they might act as a reservoir of cancer cells that spread to other parts of the body.

Using stem cells in plants

Plants have regions of rapid cell division called **meristems**. These growth regions contain stem cells that can be used to produce clones cheaply and quickly.

Meristems can be used for:
- growing and preserving rare varieties to protect them from extinction
- producing large numbers of disease-resistant crop plants.

SUMMARY

- Multicellular organisms need to have a coordinated system of structures so they can carry out vital processes, e.g. respiration, excretion, nutrition, etc.
- Differentiation is the process whereby cells become specialised.
- Stem cells have the potential to become almost any kind of cell.
- There are benefits, risks and objections to the use of stem cells.

QUESTIONS

QUICK TEST

1. Which type of specialised animal cell helps protect the lungs against dust and microorganisms?

2. Which type of specialised plant cell is used to transport water through the stem, roots and leaves?

3. Name the regions of rapid cell division in plants.

EXAM PRACTICE

1. Name the process by which cells become specialised to carry out a particular function. **[1 mark]**

2. A fertilised egg cell is called a zygote. It is a type of stem cell.

   a) The zygote will divide and then produce specialised cells. Some of these cells will become neurones. Describe **one** adaptation of a neurone. **[1 mark]**

   b) Describe **two** applications of stem cell research. **[2 marks]**

   c) Explain why some people have objections to stem cell research. **[2 marks]**

DAY 1 — 40 Minutes

Microscopy and microorganisms

Microscopes

Microscopes:
- observe objects that are too small to see with the naked eye
- are useful for showing detail at cellular and sub-cellular level.

There are two main types of microscope: the **light microscope** and the **electron microscope**.

You will probably use a light microscope in your school.

The electron microscope was invented in 1931. It has increased our understanding of sub-cellular structures because it has much higher magnifications and resolution than a light microscope.

These are white blood cells, as seen through a transmission electron microscope. The black structures are nuclei.

Comparing light and electron microscopes

Light microscope	Electron microscope
Uses light waves to produce images	Uses electrons to produce images
Low resolution	High resolution
Magnification up to ×1500	Magnification up to ×500 000 (2D) and ×100 000 (3D)
Able to observe cells and larger organelles	Able to observe small organelles
2D images only	2D and 3D images produced

HT Here are some common units used in microscopy.

Measure	Scale	Symbol
1 metre		m
1 millimetre	$\frac{1}{1\,000}$ (a thousandth) of a metre ($\times 10^{-3}$)	mm
1 micrometre	$\frac{1}{1\,000\,000}$ (a millionth) of a metre ($\times 10^{-6}$)	µm
1 nanometre	$\frac{1}{1\,000\,000\,000}$ (a thousand millionth or a billionth) of a metre ($\times 10^{-9}$)	nm
1 picometre	$\frac{1}{1\,000\,000\,000\,000}$ (a million millionth or a trillionth) of a metre ($\times 10^{-12}$)	pm

Magnification and resolution

Magnification measures how many times an object under a microscope has been made larger.

You may be asked to carry out calculations using magnification. To calculate the magnifying power of a microscope, use this formula:

$$\mathbf{magnification} = \frac{\textbf{Size of image}}{\textbf{Size of real object}}$$

Example

A light microscope produces an image of a cell which has a diameter of 1500 µm. The cell's actual diameter is 50 µm. Calculate the magnifying power of the microscope.

$$\text{magnification} = \frac{1500}{50} = \times 30$$

Resolution is the smallest distance between two points on a specimen that can still be told apart.

The diagram below shows the limits of resolution for a light microscope. When looking through it you can see fine detail down to 200 nanometres. Electron microscopes can see detail down to 0.05 nanometres!

200 nm

Staining techniques are used in light and electron microscopy to make organelles more visible.

- **Methylene blue** is used to stain the nuclei of animal cells for viewing under the light microscope.
- **Heavy metals** such as cadmium can be used to stain specimens for viewing under the electron microscope.

Culturing microorganisms

To view microorganisms under the microscope, pure, uncontaminated samples are grown or **cultured** in the laboratory. These cultures can be used for research, e.g. investigating the action of disinfectants or antibiotics.

In order to grow, microorganisms need:

- **food** in the form of a growth medium; in the laboratory, this is usually nutrient agar or broth
- **warmth** – a temperature of 25°C is used in school laboratories, as it lowers the possibility of **pathogens** growing
- **moisture**.

Aseptic technique

Aseptic technique is needed to culture microorganisms in the laboratory so that uncontaminated cultures are produced, which reduces the risk of scientists being infected by pathogens.

Bacterial growth

Under ideal conditions, a bacterial culture will grow quickly over a period of days by **binary fission**. In this process, a cell divides into two in as little time as twenty minutes!

The bacteria continue to divide in this way: 1, 2, 4, 8, 16, 32, 64, etc. So after a day, the numbers are in their millions. As they grow, the individual cells merge to produce circular **colonies** that are visible to the naked eye. You can count them or calculate their area to estimate the growth rate.

HT Calculating the area of a bacterial colony

Bacterial colonies grow **radially**, i.e. in circles. To calculate a rate of growth, the area of the colony can be worked out at two points in time.

Example: A colony of *M. luteus* is observed growing on agar in a petri dish. The diameter is measured with a ruler and is 20 mm. Exactly 24 hours later, the colony's diameter is 30 mm. What is its growth rate?

- Area of colony for first measurement $= \pi r^2$ where $\pi = 3.14$ and $r = \frac{1}{2} \times 20$
  Area $= 3.14 \times 10^2 = 314 \, mm^2$
- Area of colony for second measurement $= 3.14 \times 15^2 = 706.5 \, mm^2$
- Rate of growth $= \dfrac{706.5 - 314}{24}$
  $= \mathbf{16.35 \, mm^2 \, h^{-1}}$

QUESTIONS

QUICK TEST

1. Name the two types of microscope.
2. What three factors do microorganisms need in order to grow?

EXAM PRACTICE

1. A student grows some bacterial colonies on an agar plate. They measure the radius of eight colonies. They then use each radius to calculate the cross-sectional area of seven of the colonies.

   The results are shown in the table.

Colony	1	2	3	4	5	6	7	8
Radius/mm	10	8	2	3	2	7	5	4
Cross-sectional area/mm^2	314.0	201.0	12.6	28.3	12.6	153.9	78.5	

   a) Using the formula πr^2, calculate the cross-sectional area for colony number 8. [2 marks]
   b) Calculate the **mean** cross-sectional area for all eight colonies. [2 marks]

2. a) A student wishes to see a specimen at a higher power magnification. Which part of the light microscope must she adjust to do this? [1 mark]
   b) Explain why the student will not be able to observe the finer detail of organelles such as mitochondria. [2 marks]

SUMMARY

- **There are two main types of microscope: light microscope and electron microscope.**
- **Magnification measures how many times an object under a microscope has been enlarged.**
- **To view microorganisms under a microscope, samples are cultivated using aseptic technique.**

Cell division

Chromosomes

Chromosomes are found in the nucleus of eukaryotic cells. They are made of DNA and carry a large number of genes. Chromosomes exist as pairs called **homologues**.

For cells to duplicate exactly, it is important that all of the genetic material is duplicated. Chromosomes take part in a sequence of events that ensures the genetic code is transmitted precisely and appears in the new **daughter cells**.

DNA replication

During the cell cycle, the genetic material (made of the **polymer** molecule, DNA) is doubled and then divided between the identical daughter cells. The process of doubling is called **DNA replication**.

Mitosis

Mitosis is where a **diploid** cell (one that has a complete set of chromosomes) divides to produce two more diploid cells that are genetically identical. Most cells in the body are diploid.

Humans have a diploid number of 46.

Mitosis produces new cells:
- for growth
- to repair damaged tissue
- to replace old cells
- for asexual reproduction.

Before the cell divides, the DNA is duplicated and other organelles replicate, e.g. mitochondria and ribosomes. This ensures that there is an exact copy of all the cell's content.

Mitosis – the cell copies itself to produce two genetically identical cells

Parent cell with two pairs of chromosomes.

Each chromosome replicates (copies) itself.

Each 'daughter' cell has the same number of chromosomes, and contains the same genes, as the parent cell.

Chromosomes line up along the centre of the cell, separate into chromatids and move to opposite poles.

Meiosis

Meiosis takes place in the testes and ovaries of sexually reproducing organisms and produces gametes (eggs or sperm). The gametes are called **haploid** cells because they contain half the number of chromosomes as a diploid cell. This chromosome number is restored during **fertilisation**.

Humans have a haploid number of 23.

Meiosis – the cell divides twice to produce four cells with genetically different sets of chromosomes

Cell with two pairs of chromosomes (diploid cell).

Each chromosome replicates itself.

Cell divides for the first time.

Chromosomes part company and move to opposite poles.

Copies now separate and the second cell division takes place.

Four haploid cells (gametes), each with half the number of chromosomes of the parent cell.

Cancer

Cancer is a non-communicable disease caused by mutations in living cells.

Cancerous cells:
- divide in an uncontrolled way
- form tumours.

Benign tumours do not spread from the original site of cancer in the body. **Malignant tumour cells** invade neighbouring tissues. They spread to other parts of the body and form **secondary tumours**.

Making healthy lifestyle choices is one way to reduce the likelihood of cancer. These include:
- not smoking tobacco products (cigarettes, cigars, etc.)
- not drinking too much alcohol (causes cancer of the liver, gut and mouth)
- avoiding exposure to UV rays (e.g. sunbathing, tanning salons)
- eating a healthy diet (high fibre reduces the risk of bowel cancer) and doing moderate exercise to reduce the risk of obesity.

SUMMARY

- Multicellular organisms grow and reproduce using cell division and cell enlargement.
- Plants can also grow via differentiation into leaves, branches, etc.
- Cell division begins from the moment of fertilisation when the zygote replicates itself exactly through mitosis. Later in life, an organism may use cell division to produce sex cells (gametes) in a different type of division called meiosis.
- The different stages of cell division make up the cell cycle of an organism.

QUESTIONS

QUICK TEST

1. What name is given to sex cells?

2. Name the 46 structures in the human nucleus that carry genetic information.

3. How many chromosomes are in a human haploid cell?

EXAM PRACTICE

1. Amelia is looking at some cells through a microscope. The cells resulted from one original parent cell. The diagram below is a drawing of what she can see.

   a) Which type of cell division is shown here?   [1 mark]

   b) Explain your answer to part a).   [2 marks]

Metabolism – respiration

Metabolism

Metabolism is the sum of all the chemical reactions that take place in the body.

The two types of metabolic reaction are:
- building reactions (**anabolic**)
- breaking-down reactions (**catabolic**).

Anabolic Reactions

Anabolic reactions require the input of energy. Examples include:
- converting glucose to starch in plants, or glucose to glycogen in animals
- the synthesis of lipid molecules

fatty acids + glycerol ⟶ lipids

- the formation of **amino acids** in plants (from glucose and nitrate ions) which, in turn, are built up into proteins.

Catabolic reactions

Catabolic reactions release energy. Examples include:
- breaking down amino acids to form **urea**, which is then excreted
- respiration.

Catabolic reactions release waste energy in the form of heat (an **exothermic** reaction), which is transferred to the environment.

Respiration

Respiration continuously takes place in all organisms – the need to release energy is an essential life process. The reaction gives out energy and is therefore **exothermic**.

Aerobic respiration

Aerobic respiration takes place in cells. Oxygen and glucose molecules react and release energy. This energy is stored in a molecule called **ATP**.

glucose + oxygen ⟶ carbon dioxide + water + energy (locked in ATP)

(HT) The symbol equation for aerobic respiration is:

$$C_6H_{12}O_6 + 6O_2 \longrightarrow 6CO_2 + 6H_2O + \text{energy released}$$

Energy is used in the body for many processes, including:
- muscle contraction (for movement)
- active transport
- transmitting nerve impulses
- synthesising new molecules
- maintaining a constant body temperature.

Anaerobic respiration

Anaerobic respiration takes place in the absence of oxygen and is common in muscle cells. It quickly releases a **smaller** amount of energy than aerobic respiration through the **incomplete breakdown** of glucose.

glucose ⟶ lactic acid + energy released

In plant and yeast cells, anaerobic respiration produces different products.

glucose ⟶ ethanol + carbon dioxide + energy released

(HT) The symbol equation for anaerobic respiration in plant and yeast cells is:

$$C_6H_{12}O_6 \longrightarrow 2C_2H_5OH + 2CO_2 + \text{energy released}$$

This reaction is used extensively in the brewing and wine-making industries. It is also the initial process in the manufacture of spirits in a distillery.

Response to exercise

In animals, anaerobic respiration takes place when muscles are working so hard that the lungs and circulatory system cannot deliver enough oxygen to break down all the available glucose through aerobic respiration. In these circumstances the **energy demand** of the muscles is high.

Anaerobic respiration and recovery

Anaerobic respiration releases energy much faster over short periods of time. It is useful when short, intense bursts of energy are required, e.g. a 100 m sprint.

However, the incomplete oxidation of glucose causes **lactic acid** to build up. Lactic acid is toxic and can cause pain, cramp and a sensation of fatigue.

The lactic acid must be broken down quickly and removed to avoid cell damage and prolonged muscle fatigue.

- During exercise the body's heart rate, breathing rate and breath volume increase so that sufficient oxygen and glucose is supplied to the muscles, and so that lactic acid can be removed.
- This continues after exercise when deep breathing or panting occurs until all the lactic acid is removed. This repayment of oxygen is called **oxygen debt**.

> **HT**
> - Lactic acid is transported to the liver where it is converted back to glucose.
> - Oxygen debt is the amount of **extra** oxygen that the body needs after exercise to react with the lactic acid and remove it from the cells.

> **WS** You may be asked to present data as graphs, tables, bar charts or histograms. For example, you could show data of breathing and heart rates on a line graph.

What does the line graph below tell you about breathing and pulse rates during recovery? Why is a line graph a good way to present the data?

QUESTIONS

QUICK TEST

1. Which type of respiration releases most energy – aerobic or anaerobic?

2. What is an exothermic reaction?

3. Give the product of anaerobic respiration in humans.

4. What type of reaction is respiration – anabolic or catabolic?

EXAM PRACTICE

1. The human muscle cell and a yeast cell can each carry out anaerobic respiration.

   Both produce a toxic waste product.

   a) Write down the word equations for both human anaerobic respiration and yeast anaerobic respiration. **[2 marks]**

   b) Compare and contrast the two types of respiration in terms of how they deal with their toxic waste products. **[3 marks]**

Metabolism – enzymes

Enzyme facts

Enzymes:

- are specific, i.e. one enzyme catalyses one reaction
- have an active site, which is formed by the precise folding of the enzyme molecule
- can be denatured by high temperatures and extreme changes in pH
- have an optimum temperature at which they work – for many enzymes this is approximately 37°C (body temperature)
- have an optimum pH at which they work – this varies with the site of enzyme activity, e.g. pepsin works in the stomach and has an optimum pH of 1.5 (acidic), salivary amylase works best at pH 7.3 (alkaline).

Enzyme activity

Enzyme molecules work by colliding with **substrate** molecules and forcing them to break up or to join with others in synthesis reactions. The theory of how this works is called the **lock and key theory**.

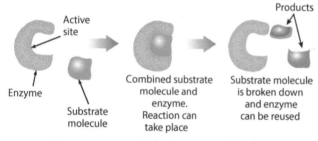

Active site

Enzyme

Substrate molecule

Combined substrate molecule and enzyme. Reaction can take place

Products

Substrate molecule is broken down and enzyme can be reused

High temperatures denature enzymes because excessive heat vibrates the atoms in the protein molecule, putting a strain on bonds and breaking them. This changes the shape of the active site.

Enzyme

Heat

Enzyme destroyed by heat

In a similar way, an extreme pH alters the active site's shape and prevents it from functioning.

Enzyme

Substrate molecule

pH level altered

Active site

Active site changed irreversibly

At lower than the optimal temperature, an enzyme still works but much more slowly. This is because the low **kinetic energy** of the substrate and enzyme molecules lowers the number of collisions that take place. When they do collide, the energy is not always sufficient to create a bond between them.

Optimum enzyme activity

Enzyme activity

Optimum temperature

Temperature

Enzymes in the digestive system

Enzymes in the digestive system help break down large nutrient molecules into smaller ones so they can be absorbed into the blood across the wall of the small intestine.

Enzyme types in the digestive system include carbohydrases, proteases and lipases.

Carbohydrases break down carbohydrates. **Amylase** is a carbohydrase that is produced in the mouth and small intestine to break down starch.

> **starch ⟶ maltose**

Other carbohydrases break down complex sugars into smaller sugars.

Proteases break down protein. The enzyme **pepsin** is a protease that is produced in the stomach.

> **protein ⟶ peptides ⟶ amino acids**

Other enzymes in the small intestine complete protein breakdown with the production of amino acids.

Lipases, which are produced in the small intestine, break down lipids.

> **lipids ⟶ fatty acids ⟶ glycerol**

Bile

Bile is a digestive liquid. It is produced in the liver and stored in the gall bladder.

- Its alkaline pH neutralises hydrochloric acid that has been produced in the stomach.
- It emulsifies fat, breaking it into small droplets with a large surface area.
- Its action enables lipase to break down fat more efficiently.

What happens to digested food?

Calorimetry

Enzyme-controlled metabolic reactions such as respiration release energy. The amount of energy contained in a substrate can be measured using **calorimetry**.

The substrate (food) is placed in a calorimeter and burned in pure oxygen. The energy released heats up the surrounding water and the temperature rise is measured.

The energy released per gram can be calculated.

$$\text{energy (J)} \longrightarrow \text{mass of water heated (g)} \times 4.2 \times \text{temperature rise (°C)}$$

WS During your course you will investigate how certain factors affect the rate of enzyme activity. These include temperature, pH and substrate concentration.

Design an investigation to discover how temperature affects the activity of amylase. Here are some guidelines.

- Iodine solution turns from a red-brown colour to blue-black in the presence of starch.
- You can measure amylase activity by timing how long it takes for iodine solution to stop turning blue-black.
- A water bath can be set up with a thermometer. Add cold or hot water to regulate the temperature.
- Identify the independent, dependent and control variables in the investigation. Write out your method in clear steps.

SUMMARY

- Enzymes are large proteins that act as biological catalysts.
- They speed up chemical reactions, including reactions that take place in living cells, e.g. respiration, photosynthesis and protein synthesis.
- Enzymes function according to the 'lock and key' theory.

QUESTIONS

QUICK TEST

1. Which two factors affect the rate of enzyme activity?

2. What type of enzyme breaks down lipids?

3. Which molecules act as building blocks for protein polymers?

EXAM PRACTICE

1. Enzymes can only function in optimum pH and temperature conditions.

   a) Explain how a temperature of 45°C will affect the shape and function of the enzyme amylase when digesting starch. **[3 marks]**

   b) Explain how a pH of 2 allows the stomach protease enzyme to function efficiently.

      Use 'lock and key' theory to aid your explanation. **[3 marks]**

Cell transport

Diffusion

Living cells need to obtain oxygen, glucose, water, mineral ions and other dissolved substances from their surroundings. They also need to excrete waste products, such as carbon dioxide or urea. These substances pass through the cell membrane by **diffusion**.

Diffusion:

- is the (net) movement of particles in a liquid or gas from a region of high concentration to one of low concentration (down a **concentration gradient**)
- happens due to the random motion of particles past each other
- stops once the particles have completely spread out
- is passive, i.e. requires no input of energy
- can be increased in terms of rate by making the concentration gradient steeper, the diffusion path shorter, increasing the temperature or increasing the surface area over which the process occurs, e.g. having a folded cell membrane.

A protist called amoeba can absorb oxygen through diffusion.

Higher concentration of oxygen

Lower concentration of oxygen

You may be asked to calculate rates of diffusion using **Fick's law**.

$$\text{rate of diffusion} \propto \frac{\text{surface area} \times \text{concentration difference}}{\text{thickness of membrane}}$$

$\propto$ means proportional to

Surface area to volume ratios

A unicellular organism (such as a protist) can absorb materials by diffusion directly from the environment. This is because it has a **large surface area to volume ratio**.

However, for a large, **multicellular organism**, the diffusion path between the environment and the inner cells of the body is long. Its large size also means that the **surface area to volume ratio** is **small**.

Adaptations

Multicellular organisms therefore need transport systems and specialised structures for exchanging materials, e.g. mammalian lungs and a small intestine, fish gills, and roots and leaves in plants. These increase diffusion efficiency in animals because they have:

- a large surface area
- a thin membrane to reduce the diffusion path
- an extensive blood supply for transport (animals)
- a ventilation system for gaseous exchange, e.g. breathing in animals.

In mammals, the individual air sacs in the lungs increase their surface area by a factor of thousands. Ventilation moves air in and out of the lungs and the heart moves blood through the capillaries. This maintains the diffusion gradient. The capillary and alveolar linings are very thin, decreasing the diffusion path.

Single alveolus showing how increased surface area enables rapid diffusion of gases

Bunches of alveoli at the end of bronchioles increase surface area (capillaries are only shown for a few alveoli)

Osmosis

Osmosis is a special case of diffusion that occurs in all organisms and involves the movement of water only.

There are two ways of describing osmosis.

1. The net movement of **water** through a partially permeable membrane from a region of **low** solute concentration to one of **high** concentration.
2. The movement of water through a partially permeable membrane down a **water potential gradient**.

Osmosis:

- occurs across a **partially permeable membrane**, so solute molecules cannot pass through (only water molecules can)
- is passive (in other words, requires no input of energy)
- allows water movement into root hair cells from the soil and between cells inside the plant

- can be demonstrated and measured in plant cells using a variety of tissues, e.g. potato chips, daffodil stems.

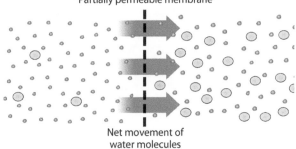

Osmosis

| Dilute solution (high water potential) | → | Concentrated solution (low water potential) |

Partially permeable membrane

Net movement of water molecules

Active transport

Substances are sometimes absorbed against a concentration gradient, i.e. from a low to a high concentration.

Active transport:
- requires the release of energy from respiration
- takes place in the small intestine in humans, where sugar is absorbed into the bloodstream
- allows plants to absorb mineral ions from the soil through root hair cells.

A cell absorbing ions by active transport

Root hair cell with high concentration of nitrate ions

Soil with lower concentration of nitrate ions

Cell uses energy to 'pull' ions in against the concentration gradient

---

WS During your course, you may investigate the effect of salt or sugar solutions on plant tissue.

Here is one experiment you could do.
1. Immerse raw potato cut into chips of equal length in sugar solutions of various concentration.
2. You will see the potato chips change in length depending on whether individual cells have lost or gained water.

Can you **predict** what would happen to the potato chips immersed in:
- concentrated sugar (e.g. 1 molar)
- medium concentration sugar (e.g. 0.5 molar)
- water (0 molar)?

SUMMARY

- Living cells obtain oxygen, glucose, water and other substances from their surroundings, and excrete waste products, through diffusion.
- Osmosis is a special type of diffusion involving only water.
- Active transport requires the release of energy from respiration.

QUESTIONS

QUICK TEST

1. What is Fick's Law used for?
2. Some plant tissue is placed in a highly concentrated salt solution. Explain why water leaves the cells.

EXAM PRACTICE

1. Root hair cells absorb ions from the soil and water that surround them.

   Explain how the ions can enter, despite the fact that this is against a concentration gradient. **[2 marks]**

2. A student places a thin piece of rhubarb epidermis in a strong sugar solution and observes the cells under a microscope.

   `She notices that the cytoplasm has pulled away from the cell wall.

   Explain this change. **[4 marks]**

Plant tissues, organs and systems

Leaves

As this cross-section of a leaf shows, leaf tissues are adapted for efficient photosynthesis. The epidermis covers the upper and lower surfaces of the leaf and protects the plant against pathogens.

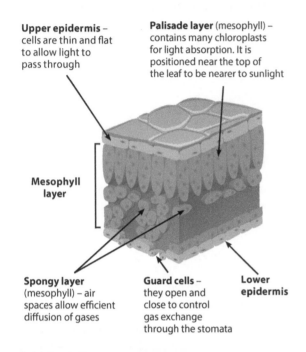

Upper epidermis – cells are thin and flat to allow light to pass through

Palisade layer (mesophyll) – contains many chloroplasts for light absorption. It is positioned near the top of the leaf to be nearer to sunlight

Mesophyll layer

Spongy layer (mesophyll) – air spaces allow efficient diffusion of gases

Guard cells – they open and close to control gas exchange through the stomata

Lower epidermis

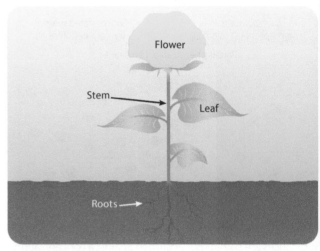

Flower

Stem

Leaf

Roots →

Stem and roots

Veins in the stem, roots and leaves contain tissues that transport water, carbohydrate and minerals around the plant.

- **Xylem tissue** transports water and mineral ions from the roots to the rest of the plant.

- **Phloem tissue** transports dissolved sugars from the leaves to the rest of the plant.

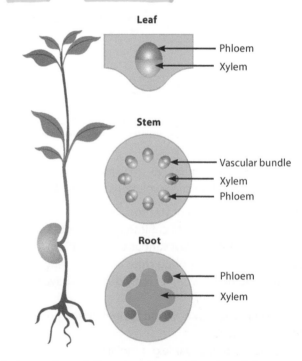

Leaf
- Phloem
- Xylem

Stem
- Vascular bundle
- Xylem
- Phloem

Root
- Phloem
- Xylem

Meristem tissue is found at the growing tips of shoots and roots.

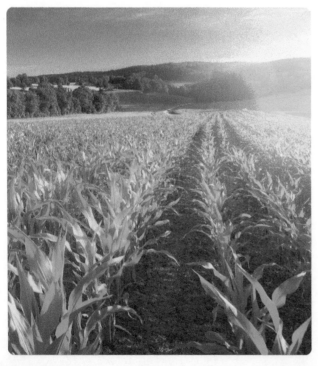

Xylem, phloem and root hair cells

Xylem, phloem and root hair cells are adapted to their function.

Part of plant	Appearance	Function	How they are adapted to their function
Xylem	Hollow tubes made from dead plant cells (the hollow centre is called a lumen)	Transport water and mineral ions from the roots to the rest of the plant in a process called transpiration	The cellulose cell walls are thickened and strengthened with a waterproof substance called lignin
Phloem	Columns of living cells	Translocate (move) cell sap containing sugars (particularly sucrose) from the leaves to the rest of the plant, where it is either used or stored	Phloem have pores in the end walls so that the cell sap can move from one phloem cell to the next
Root hair cells	Long and thin; have hair-like extensions	Absorb minerals and water from the soil	Large surface area

Xylem   Phloem

Xylem   Phloem

SUMMARY

- A plant is made up of organs and tissues that enable it to be a photosynthetic organism.
- Roots absorb water and minerals. They anchor plants in the soil.
- The stem transports water and nutrients to leaves. It holds leaves up to the light for maximum absorption of energy.
- The leaf is the organ of photosynthesis.
- The flower makes sexual reproduction possible through pollination.

QUESTIONS

QUICK TEST

1. What part of the plant is the main organ for photosynthesis?

2. What is the name of the waterproof substance that strengthens xylem?

EXAM PRACTICE

1. **a)** State one structural difference between xylem and phloem tissue. **[1 mark]**

   b) Small, herbivorous insects called aphids are found on plant stems. They have piercing mouthparts that can penetrate down to the phloem.

   Explain the reasons for this behaviour. **[2 marks]**

Transport in plants

Transpiration

The movement of water through a plant, from roots to leaves, takes place as a transpiration stream. Once water is in the leaves, it diffuses out of the stomata into the surrounding air. This is called (evapo)transpiration.

| Water evaporates from the spongy mesophyll through the stomata. | → | Water passes by osmosis from the xylem vessels in the leaf into the spongy mesophyll cells to replace what has been lost. | → | This movement 'pulls' the column of water in that xylem vessel upwards. | → | Water enters root hair cells by osmosis to replace water that has entered the xylem. |

Measuring rate of transpiration

A leafy shoot's rate of transpiration can be measured using a **potometer**.

A potometer

The shoot is held in a tube with a bung around the top to prevent any water from evaporating (this would give a false measurement of the water lost by transpiration) and to prevent air bubbles entering the apparatus.

As the plant transpires, it takes up water from the tube to replace what it has lost. All the water is then pulled up, moving the air bubble along.

The distance the air bubble moves can be used to calculate the plant's rate of transpiration for a given time period.

The experiment can be repeated, varying a different factor each time, to see how each factor affects the rate of transpiration.

Factors affecting rate of transpiration

Evaporation of water from the leaf is affected by **temperature**, **humidity**, **air movement** and **light intensity**.

- **Increased temperature** increases the kinetic energy of molecules and removes water vapour more quickly away from the leaf.
- **Increased air movement** removes water vapour molecules.
- **Increased light intensity** increases the rate of transpiration because it causes the stomata to open.
- **Decreasing atmospheric humidity** lowers water vapour concentration outside of the stoma and so maintains the concentration gradient.

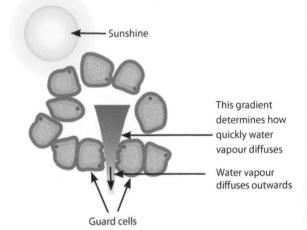

How water vapour exits the leaf

Opening and closing of stomata

Guard cells control the amount of water vapour that evaporates from the leaves and the amount of carbon dioxide that enters them.

- When light intensity is high and photosynthesis is taking place at a rapid rate, the sugar concentration rises in photosynthesising cells, e.g. palisade and guard cells.
- Guard cells respond to this by increasing the rate of water movement in the transpiration stream. This in turn provides more water for photosynthesis.

Sunshine

High light intensity. Increased photosynthesis rate

More sugar manufactured in guard cells. This raises **solute** concentration in guard cells therefore water moves into them

Guard cells have increased **turgidity**, which increases size of stoma allowing more rapid loss of water vapour

Stoma

WS A **hypothesis** is an idea or explanation that you test through study and experiments. It should include a reason. For example: desert plants have fewer stomata than temperate plants **because** they need to minimise water loss.

- In an experiment investigating the factors that affect the rate of transpiration, a student plans to take measurements of weight loss or gain from a privet plant.
- Construct hypotheses for each of these factors: **temperature**, **humidity**, **air movement** and **light intensity**.
- The first has been done for you: As temperature increases, the plant will lose mass/water more quickly **because** diffusion occurs more rapidly.

Stomata

SUMMARY

- A plant's system is made up of organs and tissues that enable it to be a photosynthetic organism.
- Roots absorb water and minerals. They anchor plants in the soil.
- The stem transports water and nutrients to leaves. It holds leaves up to the light for maximum absorption of energy.
- The leaf is the organ of photosynthesis.

QUESTIONS

QUICK TEST

1. What effect would decreasing air humidity have on transpiration?

2. Which cells control the opening and closing of the stomata?

3. What equipment is used to measure a leafy shoot's rate of transpiration?

EXAM PRACTICE

1. A student set up a plant in a potometer and noted the position of the air bubble at 2.1 cm on the scale. He left the apparatus in still air at 20°C for exactly 60 minutes. He noted the new position of the air bubble on the scale as 2.3 cm. He then immediately placed a fan close to the plant and switched it on. He noted the temperature was still 20°C. After another 60 minutes he saw that the bubble was at 5.7 cm.

   a) Calculate the rate of movement of the bubble in cm per minute for each period of time.

      Show your working. **[4 marks]**

   b) Explain the difference between both sets of results. **[3 marks]**

Transport in humans 1

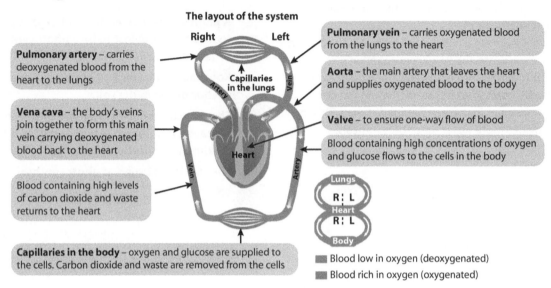

The layout of the system

Right   Left

Pulmonary artery – carries deoxygenated blood from the heart to the lungs

Vena cava – the body's veins join together to form this main vein carrying deoxygenated blood back to the heart

Blood containing high levels of carbon dioxide and waste returns to the heart

Capillaries in the lungs

Artery   Vein

Heart

Vein   Artery

Pulmonary vein – carries oxygenated blood from the lungs to the heart

Aorta – the main artery that leaves the heart and supplies oxygenated blood to the body

Valve – to ensure one-way flow of blood

Blood containing high concentrations of oxygen and glucose flows to the cells in the body

Lungs
R : L
Heart
R : L
Body

Capillaries in the body – oxygen and glucose are supplied to the cells. Carbon dioxide and waste are removed from the cells

■ Blood low in oxygen (deoxygenated)
■ Blood rich in oxygen (oxygenated)

Blood circulation

Blood moves around the body in a **double circulatory system**. In other words, blood moves twice through the heart for every full circuit. This ensures maximum efficiency for absorbing oxygen and delivering materials to all living cells.

The heart

The heart is made of powerful muscles that contract and relax rhythmically in order to continuously pump blood around the body. The **heart muscle** is supplied with nutrients (particularly glucose) and oxygen through the coronary artery.

The sequence of events that takes place when the heart beats is called the **cardiac cycle**.

- The heart relaxes and blood enters both atria from the veins.
- The atria contract together to push blood into the ventricles, opening the atrioventricular valves.
- The ventricles contract from the bottom, pushing blood upwards into the arteries. The backflow of blood into the atria is prevented by the closing of the atrioventricular valves.
- The heart relaxes again and backflow into the ventricles is prevented by the closing of the **semilunar valves**.

The left side of the heart is more muscular than the right because it has to pump blood further round the body. The right side only has to pump blood to the lungs and back.

A useful measurement for scientists and doctors to take is **cardiac output**. This is calculated using:

> **cardiac output = stroke volume × heart rate**

So, for a person who pumps out 70 ml of blood in one heartbeat (stroke volume) and has a pulse of 70 beats per minute, the cardiac output would be 4900 ml per minute.

Controlling the heartbeat

The heart is stimulated to beat rhythmically by pacemaker cells.

The pacemaker cells produce impulses that spread across the atria to make them contract. Impulses are spread from here down to the ventricles, making them contract, pushing blood up and out.

Pacemaker cells

Nerves connecting the heart to the brain can increase or decrease the pace of the pacemaker cells in order to regulate the heartbeat.

If a person has an irregular heartbeat, they can be fitted with an artificial, electrical pacemaker.

Blood vessels

Blood is carried through the body in three types of vessel:

- **Arteries** have thick walls made of elastic fibres and muscle fibres to cope with the high pressure. The **lumen** (space inside) is small compared to the thickness of the walls. There are no valves.
- **Veins** have thinner walls. The lumen is much bigger compared to the thickness of the walls and there are valves to prevent the backflow of blood.
- **Capillaries** are narrow vessels with walls only one cell thick. These microscopic vessels connect arteries to veins, forming dense networks or **beds**. They are the only blood vessels that have permeable walls to allow the exchange of materials.

Artery

Lumen

Vein

Lumen

Valve

Capillary

Note: capillaries are much smaller than veins or arteries

Coronary heart disease

Coronary heart disease is a non-communicable disease. It results from the build-up of **cholesterol**, leading to plaques laid down in the coronary arteries. This restricts blood flow and the artery may become blocked with a blood clot or **thrombosis**. The heart muscle is deprived of glucose and oxygen, which causes a **heart attack**.

The likelihood of plaque developing increases if you have a high fat diet. The risk of a heart attack can be reduced by:

- eating a balanced diet and not being overweight
- not smoking tobacco
- lowering alcohol intake
- reducing salt levels in your diet
- reducing stress levels.

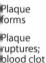
Healthy artery
Build-up of fatty material begins
Plaque forms
Plaque ruptures; blood clot forms

Coronary arteries

SUMMARY

- **Blood moves around the body in a double circulatory system.**
- **The heart has strong muscles that contract and relax to pump blood around the body.**
- **There are three types of blood vessels: arteries, veins and capillaries.**
- **Coronary heart disease can occur due to build-up of cholesterol in arteries.**

QUESTIONS

QUICK TEST

1. Which blood vessels contain valves?
2. Capillaries are smaller than veins. True or false?
3. Which vessel carries oxygenated blood from the lungs to the heart?

EXAM PRACTICE

1. a) Explain why arteries have thick, elastic muscle walls. **[1 mark]**
   b) Why do veins have valves? **[1 mark]**
2. Compared with reptiles and amphibians, humans have a double circulation.
   a) What is meant by this term? **[1 mark]**
   b) What is the advantage of having a double circulation? **[1 mark]**

Transport in humans 2

Remedying heart disease

For patients who have heart disease, artificial implants called **stents** can be used to increase blood flow through the coronary artery.

Statins are a type of drug that can be taken to reduce blood cholesterol levels.

In some people, the heart valves may deteriorate, preventing them from opening properly. Alternatively, the valve may develop a leak.

This means that the supply of oxygenated blood to vital organs is reduced. The problem can be corrected by surgical replacement using a **biological** or **mechanical** valve.

When complete heart failure occurs, a heart transplant can be carried out. If a donor heart is unavailable, the patient may be kept alive by an artificial heart until one can be found. Mechanical hearts are also used to give the biological heart a rest while it recovers.

Blood as a tissue

Blood transports digested food and oxygen to cells and removes the cells' waste products. It also forms part of the body's defence mechanism.

The four components of blood are:
- platelets
- plasma
- white blood cells
- red blood cells.

Platelets
Tiny cell fragments. They initiate the clotting process and repair damage to blood vessels.

Plasma
A straw-coloured liquid that transports dissolved food materials, urea, carbon dioxide, protein, hormones and antibodies.

White blood cells (phagocytes and lymphocytes) Protect the body against pathogens. Some have a flexible shape, which enables them to engulf invading microorganisms. Others produce antibodies.

Red blood cells (erythrocytes) Transport oxygen from lungs to tissues.

Oxygen transport

Red blood cells are small and have a biconcave shape. This gives them a large surface area to volume ratio for absorbing oxygen. When the cells reach the lungs, they absorb oxygen using a red protein called haemoglobin.

> **haemoglobin + oxygen ⇌ oxyhaemoglobin**

Blood is then pumped around the body to the tissues, where the reverse of the reaction takes place. Oxygen diffuses out of the red blood cells and into the tissues.

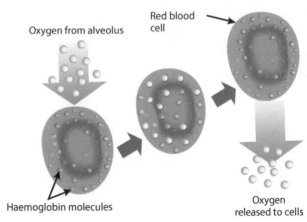

Transport of oxygen in red blood cells

Oxygen from alveolus

Red blood cell

Haemoglobin molecules

Oxygen released to cells

The lungs

Humans, like many vertebrates, have lungs to act as a **gaseous exchange surface**.

Other structures in the thorax enable air to enter and leave the lungs (ventilation).

- The trachea is a flexible tube, surrounded by rings of cartilage to stop it collapsing. Air is breathed in via the mouth and passes through here on its way to the lungs.
- Bronchi are branches of the trachea.
- The alveoli are small air sacs that provide a large surface area for the exchange of gases.
- Capillaries form a dense network to absorb maximum oxygen and release carbon dioxide.

In the alveoli, **oxygen** diffuses down a concentration gradient. It moves across the thin layers of cells in the alveolar and capillary walls, and into the red blood cells.

For **carbon dioxide**, the gradient operates in reverse. The carbon dioxide passes from the blood to the alveoli, and from there it travels back up the air passages to the mouth.

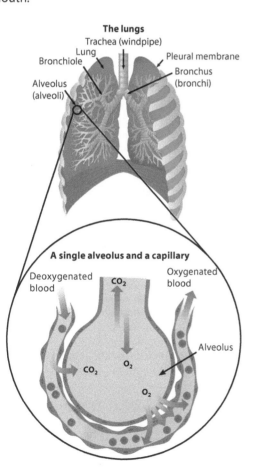

The lungs

Trachea (windpipe)

Lung

Bronchiole

Pleural membrane

Bronchus (bronchi)

Alveolus (alveoli)

A single alveolus and a capillary

Deoxygenated blood

CO_2

Oxygenated blood

CO_2

O_2

Alveolus

O_2

SUMMARY

- **There are four components of blood: platelets, plasma, white blood cells and red blood cells.**
- **Oxygen is transported around the body by red blood cells.**
- **The lungs act as a gaseous exchange surface.**

QUESTIONS

QUICK TEST

1. Name the four components of blood.

2. What is produced when haemoglobin combines with oxygen?

EXAM PRACTICE

1. Describe the function of white blood cells. **[2 marks]**

2. A smoker's lungs develop a layer of tar lining the inside of the alveoli.

   Explain why this will make the smoker breathless.

   Use ideas about diffusion in your explanation. **[3 marks]**

Photosynthesis

Photosynthesis:

- is an endothermic reaction
- requires chlorophyll to absorb the sunlight; this is found in the **chloroplasts** of photosynthesising cells, e.g. palisade cells, guard cells and spongy mesophyll cells
- produces **glucose**, which is then respired for energy release or converted to other useful molecules for the plant
- produces **oxygen** that has built up in the atmosphere over millions of years; oxygen is vital for respiration in all organisms.

carbon dioxide + water $\xrightarrow[\text{chloroplast}]{\text{light energy}}$ glucose + oxygen

HT $6CO_2 + 6H_2O \xrightarrow[\text{chloroplast}]{\text{light energy}} C_6H_{12}O_6 + 6O_2$

Rate of photosynthesis

The rate of photosynthesis can be affected by:

- temperature
- light intensity
- carbon dioxide concentration
- amount of chlorophyll.

In a given set of circumstances, **temperature**, **light intensity** and **carbon dioxide concentration** can act as limiting factors.

Temperature

1. As the temperature rises, so does the rate of photosynthesis. This means temperature is limiting the rate of photosynthesis.
2. As the temperature approaches 45°C, the enzymes controlling photosynthesis start to be denatured. The rate of photosynthesis decreases and eventually declines to zero.

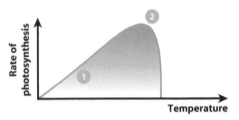

Light intensity

1. As the light intensity increases, so does the rate of photosynthesis. This means light intensity is limiting the rate of photosynthesis.
2. Eventually, the rise in light intensity has no effect on photosynthesis rate. Light intensity is no longer the limiting factor; carbon dioxide or temperature must be.

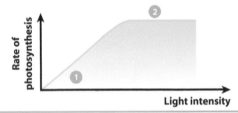

Carbon dioxide concentration

1. As carbon dioxide concentration increases, so does the rate of photosynthesis. Carbon dioxide concentration is the limiting factor.
2. Eventually, the rise in carbon dioxide concentration has no effect – it is no longer the limiting factor.

You need to understand that each factor has the potential to increase the rate of photosynthesis.

HT You also need to explain how these factors interact in terms of which variable is acting as the limiting factor.

HT The inverse law

The effect of light intensity on photosynthesis can be investigated by placing a lamp at varying distances from a plant. As the lamp is moved further away from the plant the light intensity decreases, as shown in the graph.

Distance from plant

There is an **inverse relationship** between the two variables. The graph can be used to convert distances to light intensity, or light intensity can be calculated using the formula:

$$\text{light intensity} = \frac{1}{d^2}$$

- d is the distance from the lamp.

This formula produces a dimensionless quantity of light intensity, so no units need to be stated. Instruments that measure absolute light intensity may use units such as 'lux' or the 'candela'.

Commercial applications

Farmers and market gardeners can increase their crop yields in greenhouses. They do this by:

- making the temperature optimum for growth using heaters
- increasing light intensity using lamps
- installing fossil-fuel burning stoves to increase carbon dioxide concentration (and increase temperature).

If applied carefully, the cost of adding these features will be offset by increased profit from the resulting crop.

SUMMARY

- **Plants can photosynthesise, i.e. make food molecules in the form of carbohydrate from carbon dioxide and water. As such, they are the main producers of biomass.**
- **Sunlight energy is needed for photosynthesis.**
- **The rate of photosynthesis is affected by light intensity, temperature and carbon dioxide concentration.**
- **Other products such as starch and cellulose are made from glucose.**

Uses of glucose in plants

The glucose produced from photosynthesis can be used immediately in respiration, but some is used to synthesise larger molecules: starch, cellulose, protein and lipids.

Starch is insoluble. So it is suitable for storage in leaves, stem or roots.

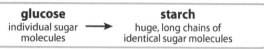

Cellulose is needed for cell walls.

Protein is used for the growth and repair of plant tissue, and also to synthesise enzyme molecules.

Lipids are needed in cell membranes, and for fat and oil storage in seeds.

QUESTIONS

QUICK TEST

1. Give two reasons why photosynthesis is seen as the opposite of respiration.
2. What is cellulose needed for?

EXAM PRACTICE

HT 1. The graph shows how the rate of photosynthesis changes in a plant with changing light intensity.

A market gardener understands this and wants to use the information to increase the yield of her tomatoes.

a) Use the graph to explain how she could accomplish this. **[4 marks]**

b) Explain why she should not invest too much money in her solution. **[2 marks]**

Non-communicable diseases

Diseases

Communicable diseases are caused by pathogens such as bacteria and viruses. They can be transmitted from organism to organism in a variety of ways. Examples include cholera and tuberculosis.

Non-communicable diseases are not primarily caused by pathogens. Examples are diseases caused by a poor diet, diabetes, heart disease and smoking-related diseases.

Health is the state of physical, social and mental well-being. Many factors can have an effect on health, including stress and life situations.

Risk factors

Non-communicable diseases often result from a combination of several **risk factors**.

- Risk factors produce an increased likelihood of developing that particular disease. They can be aspects of a person's lifestyle or substances found in the body or environment.
- Some of these factors are difficult to quantify or to establish as a definite **causal connection**. So scientists have to describe their effects in terms of probability or likelihood.

The symptoms observed in the body may result from communicable and non-communicable components interacting.

A lowered immune system may make a person more vulnerable to infection.

Immune reactions caused by pathogens can trigger allergies such as asthma and skin rashes.

Symptoms

Viruses inhabiting living cells can change them into cancer cells.

Serious physical health problems can lead to **mental illness** such as depression.

Poor diet

People need a **balanced diet**. If a diet does not have the correct amount of the main food groups, malnutrition

might result. Lack of correct vitamins leads to diseases such as **scurvy** and **rickets**. Lack of the mineral, iron, results in **anaemia**. A high fat diet contributes to cardiovascular disease and high levels of salt increase blood pressure.

Smoking tobacco

Chemicals in tobacco smoke affect health.

- **Carbon monoxide** decreases the blood's oxygen-carrying capacity.
- **Nicotine** raises the heart rate and therefore blood pressure.
- **Tar** triggers cancer.
- **Particulates** cause **emphysema** and increase the likelihood of **lung infections**.

Weight/lack of exercise

Obesity and lack of exercise both increase the risk of developing **type 2 diabetes** and cardiovascular disease.

One way to show whether someone is underweight or overweight for their height is to calculate their **body mass index** (**BMI**), using the following formula:

$$BMI = \frac{mass\ (kg)}{height\ (m)^2}$$

Recommended BMI chart

BMI	What it means
<18.5	Underweight – too light for your height
18.5–25	Ideal – correct weight range for your height
25–30	Overweight – too heavy
30–40	Obese – much too heavy. Health risks!

Example

Calculate a man's BMI if he is 1.65 m tall and weighs 68 kg.

$$BMI = \frac{mass\ (kg)}{height\ (m)^2} = \frac{68}{1.65^2} = \frac{68}{2.7} = \textbf{25}$$

The recommended BMI for his height (1.65 m) is 18.5–25, so he is just a healthy weight.

There are drawbacks to using BMI as a way of assessing people's health. For example:

- teenagers go through a rapid growth phase

● a person could have a well-developed muscle system – this would increase their body mass but not make them obese.

Some scientists say a more accurate method is using the waist/hip ratio. A tape measure is used to measure the circumference of the hips and the waist (at its widest). The waist measurement is divided by the hips measurement. The following chart can then be used.

Waist to hip ratio (WHR)		
Male	**Female**	**Health risk based solely on WHR**
0.95 or below	0.80 or below	Low
0.96 to 1.0	0.81 to 0.85	Moderate
1.0+	0.85+	High

Alcohol

Drinking excess alcohol can impair brain function and lead to **cirrhosis** of the liver. It also contributes to some types of cancer and cardiovascular disease.

Smoking and drinking alcohol during pregnancy

Unborn babies receive nutrition from the mother via the placenta. Substances from tobacco, alcohol and other drugs can pass to the baby and cause **lower birth weight**, **foetal alcohol syndrome** and **addiction**.

Carcinogens

Exposure to **ionising radiation** (e.g. X-rays, gamma rays) can cause cancerous tumours. Overexposure to UV light can cause skin cancer. Certain chemicals such as mercury can also increase the likelihood of cancer.

SUMMARY

● **Diseases can be communicable or non-communicable.**

● **Communicable diseases are caused by pathogens. Non-communicable diseases include heart disease, diabetes and smoking-related illnesses.**

● **Symptoms of diseases are often due to an interaction between communicable and non-communicable factors.**

● **Risk factors include poor diet, smoking tobacco, lack of exercise and alcohol abuse.**

(WS) Interpreting complex data in graphs doesn't need to be difficult. This line graph shows data about smoking and lung cancer. Look for different patterns in it. For example:

● males have higher smoking rates in all years

● female cancer rates have increased overall since 1972.

Can you see any other patterns?

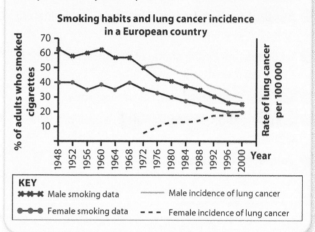

Smoking habits and lung cancer incidence in a European country

KEY
✖✖✖ Male smoking data
●●● Female smoking data
—— Male incidence of lung cancer
--- Female incidence of lung cancer

QUESTIONS

QUICK TEST

1. What does BMI stand for?

2. What is a pathogen?

EXAM PRACTICE

1. Write down two conditions that are non-communicable. **[2 marks]**

2. a) A woman's height is 1.55 m and she weighs 64 kg. What is her BMI? **[2 marks]**

   b) Using the BMI table on p.28, state whether you think she should take steps to lose weight.

      Give a reason for your answer. **[2 marks]**

Communicable diseases

How do pathogens spread?

Pathogens are disease-causing microorganisms from groups of bacteria, viruses, fungi and protists. All animals and plants can be affected by pathogens. They spread in many ways, including:

- **droplet infection** (sneezing and coughing), e.g. flu
- **physical contact**, such as touching a contaminated object or person, e.g. Ebola virus
- **transmission** by transferral of or contact with bodily fluids, e.g. hepatitis B
- **sexual transmission**, e.g. HIV, gonorrhoea
- **contamination of food or water**, e.g. Salmonella, cholera
- **animal bites**, e.g. rabies.

How do pathogens cause harm?

- Bacteria and viruses reproduce rapidly in the body.
- Viruses cause cell damage.
- Bacteria produce toxins that damage tissues.

These effects produce **symptoms** in the body.

How can the spread of disease be prevented?

The spread of disease can be prevented by:

- good hygiene, e.g. washing hands/whole body, using soaps and disinfectants
- destroying vectors, e.g. disrupting the life cycle of mosquitoes can combat malaria
- the isolation or quarantine of individuals
- vaccination.

Bacterial diseases

Disease	Transmission	Symptoms	Treatment/prevention
Tuberculosis	Droplet infection	Persistent coughing, which may bring up blood; chest pain; weight loss; fatigue; fever; night sweats; chills	Long course of antibiotics
Cholera	Contaminated water/food	Diarrhoea; vomiting; dehydration	Rehydration salts
Chlamydia	Sexually transmitted	May not be present, but can include discharge and bleeding from sex organs	Antibiotics Using condoms during sexual intercourse can reduce chances of infection
Helicobacter	Spread orally from person to person by saliva, or by fecal contamination	Abdominal pain; feeling bloated; nausea; vomiting; loss of appetite; weight loss	Proton pump inhibitors and antibiotics
Salmonella	Contaminated food containing toxins from pathogens – these could be introduced from unhygienic food preparation techniques	Vomiting; fever; diarrhoea; stomach cramps	Anti-diarrhoeals and antibiotics; vaccinations for chickens
Gonorrhoea	Sexually transmitted	Thick yellow or green discharge from vagina or penis; pain on urination	Antibiotic injection followed by antibiotic tablets; penicillin is no longer effective against gonorrhoea; prevention through use of condoms

Fungal diseases

Disease	Transmission	Symptoms	Treatment/prevention
Athlete's foot	Direct and indirect contact, e.g. skin-to-skin, bed sheets and towels (often spreads at swimming pools and in changing rooms)	Itchy, red, scaly, flaky and dry skin	Self-care and anti-fungal medication externally applied

Viral and protist diseases

Disease	Transmission	Symptoms and notes	Treatment/prevention
Measles (viral)	Droplets from sneezes and coughs	Fever; red skin rash; fatal if complications arise	No specific treatment; vaccine is a highly effective preventative measure
HIV (viral)	Sexually transmitted; exchange of body fluids; sharing of needles during drug use	Flu-like symptoms initially; late-stage AIDS produces complications due to compromised immune system	Anti-retroviral drugs
Ebola (viral)	Via body fluids; contaminated needles; bite from infected animal	Early stages: muscle pain; sore throat; diarrhoea Later stages: kidney/liver failure; internal bleeding	No direct treatments or tested vaccines are available at the time of writing; symptoms and infections are treated as they appear
Malaria (protist)	Via mosquito vector	Headache; sweats; chills and vomiting; symptoms disappear and reappear on a cyclical basis; further life-threatening complications may arise	Various anti-malarial drugs are available for both prevention and cure; prevention of mosquito breeding and use of mosquito nets

Viruses carry out their life cycle largely within a host cell. Viruses tend to follow one of two types of cycle.

- **Lysogenic**: the virus does not lyse (split) the host cell. Instead, the viral DNA becomes part of the host DNA. When the host cell divides, the viral DNA is also copied and is present in each of the daughter cells.

- **Lytic**: the virus replicates using the host DNA (as before). The new viruses that are formed destroy the infected cell and its membrane as they are released.

Influenza viral life cycle

Binding to the target cell

Budding

Replication

Viral protein synthesis

New virion

QUESTIONS

QUICK TEST

1. Why does cholera spread quickly in areas that have poor sanitation?

2. Name a fungal disease.

EXAM PRACTICE

1. Microorganisms consist of bacteria, viruses, fungi and protista.

   Many cause harm to the human body.

   a) Write down the term that describes such organisms. **[1 mark]**

   b) Harmful microorganisms produce symptoms when they reproduce in large numbers.

      Write down **two** ways in which the microorganisms cause damage. **[2 marks]**

2. Malaria kills many thousands of people every year. The disease is common in areas that have warm temperatures and stagnant water.

   Explain why this is. **[2 marks]**

SUMMARY

- **Pathogens can be spread through physical contact, droplet infection, sexual transmission, contamination of food/water, transmission of bodily fluids, or animal bites.**
- **Bacterial diseases include salmonella and gonorrhoea.**
- **Viral and protist diseases include measles, HIV and malaria.**

Human defences

Non-specific defences

The body has a number of general or non-specific defences to stop pathogens multiplying inside it.

The skin covers most of the body – it is a **physical barrier** to pathogens. It also secretes antimicrobial peptides to kill microorganisms. If the skin is damaged, a clotting mechanism takes place in the blood preventing pathogens from entering the site of the wound.

Tears contain enzymes called **lysozymes**. Lysozymes break down pathogen cells that might otherwise gain entry to the body through tear ducts.

Hairs in the nose trap particles that may contain pathogens.

Tubes in the respiratory system (**trachea** and **bronchi**) are lined with special epithelial cells. These cells either produce a sticky, liquid mucus that traps microorganisms or have tiny hairs called cilia that move the mucus up to the mouth where it is swallowed.

The stomach produces **hydrochloric acid**, which kills microorganisms.

Phagocytes are a type of white blood cell. They move around in the bloodstream and body tissues searching for pathogens. When they find pathogens, they engulf and digest them in a process called **phagocytosis**.

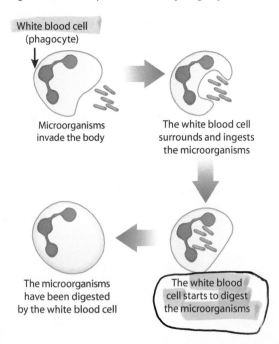

White blood cell (phagocyte)

Microorganisms invade the body

The white blood cell surrounds and ingests the microorganisms

The white blood cell starts to digest the microorganisms

The microorganisms have been digested by the white blood cell

Specific defences

White blood cells called **lymphocytes** recognise molecular markers on pathogens called antigens. They produce antibodies that lock on to the antigens on the cell surface of the pathogen cell. The immobilised cells are clumped together and engulfed by phagocytes.

White blood cell (lymphocyte)

Microorganisms (with antigens) invade the body

The white blood cell forms antibodies

A phagocyte destroys the microorganisms

The antibodies cause the microorganisms to clump

Some white blood cells produce **antitoxins** that neutralise the poisons produced from some pathogens.

Every pathogen has its own unique antigens. Lymphocytes make antibodies specifically for a particular antigen.

Example: Antibodies to fight TB will not fight cholera

TB microorganism

Antibodies

Antigens

White blood cells

Cholera microorganism

Active immunity

Once lymphocytes recognise a particular pathogen, the interaction is stored as part of the body's immunological memory through memory lymphocytes. These memory cells can produce the right antibodies much quicker if the same pathogen is detected again, therefore providing future protection against the disease. The process is called the secondary response and is part of the body's active immunity.

Active immunity can also be achieved through vaccination.

Memory lymphocytes and antibody production

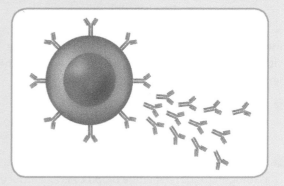

WS Investigating the growth of microorganisms involves culturing microorganisms.

This presents hazards that require a risk assessment. A risk assessment involves taking into account the severity of each hazard and the likelihood that it will occur.

Any experiment of this type involves thinking about risks in advance. Here is an example of a risk assessment table for this investigation.

Hazard	Infection from pathogen	Scald from autoclave (a specialised pressure cooker for superheating its contents)
Risk	High	High
How to lower the risk	• Observe aseptic technique. • Wash hands thoroughly before and after experiment. • Store plates at a maximum temperature of 25°C.	• Ensure lid is tightly secured. • Adjust heat to prevent too high a pressure. • Wait for autoclave to cool down before removing lid.

SUMMARY

• The body has a number of non-specific defences, including the skin, tears, and stomach acid.

• The body has specific defences in the form of lymphocytes.

• Lymphocytes produce antibodies, which lock on to antigens.

• Memory cells remember the shape of antigens and provide a defence against further infection. This is active immunity.

QUESTIONS

QUICK TEST

1. Describe the process of phagocytosis.

2. What is an antigen?

3. What enzyme is found in tears?

EXAM PRACTICE

1. The diagram shows a white blood cell producing small proteins as part of the body's immune system.

   a) What is the name of these proteins? **[1 mark]**

   b) i) What is the name of this type of white blood cell? **[1 mark]**

   ii) Describe what happens to these components if the body is invaded again by the same pathogen. **[3 marks]**

Fighting disease

Vaccination

There are two types of vaccination: active and passive.

Passive immunisation

Antibodies are introduced into an individual's body, rather than the person producing them on their own. Some pathogens or toxins (e.g. snake venom) act very quickly and a person's immune system cannot produce antibodies quickly enough. So the person must be injected with the antibodies. However, this does not give long-term protection.

Active immunisation

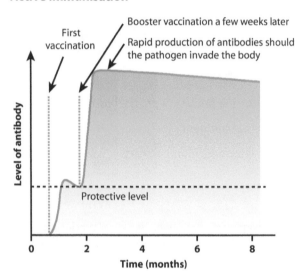

Labels on graph: First vaccination; Booster vaccination a few weeks later; Rapid production of antibodies should the pathogen invade the body; Protective level

y-axis: Level of antibody
x-axis: Time (months) — 0, 2, 4, 6, 8

Immunisation gives a person immunity to a disease without the pathogens multiplying in the body, or the person having symptoms.

- A weakened or inactive strain of the pathogen is injected. The pathogen is heat-treated so it cannot multiply. The antigen molecules remain intact.

- Even though they are harmless, the antigens on the pathogen trigger the white blood cells to produce specific antibodies.

- As with natural immunity, **memory lymphocytes** remain sensitised. This means they can produce more antibodies very quickly if the same pathogen is detected again.

HT

Benefits of immunisation	Risks of immunisation
• It protects against diseases that could kill or cause disability (e.g. polio, measles). • If everybody is vaccinated and herd immunity is established, the disease eventually dies out (this is what happened to smallpox).	• A person could have an allergic reaction to the vaccine (small risk).

Antibiotics and painkillers

Diseases caused by bacteria (not viruses) can be treated using **antibiotics**, e.g. penicillin. Antibiotics are drugs that destroy the pathogen. Some bacteria need to be treated with antibiotics specific to them.

Antibiotics work because they **inhibit** cell processes in the bacteria but not the body of the host.

Viral diseases can be treated with **antiviral drugs**, e.g. swine flu can be treated with 'Tamiflu' tablets. It is a challenge to develop drugs that destroy viruses without harming body tissues.

Antibiotic resistance

Antibiotics are very effective at killing bacteria. However, some bacteria are **naturally resistant** to particular antibiotics. It is important for patients to follow instructions carefully and take the full course of antibiotics so that all the harmful bacteria are killed.

If doctors over-prescribe antibiotics, there is more chance of resistant bacteria surviving. These multiply and spread, making the antibiotic useless. **MRSA** is a bacterium that has become resistant to most antibiotics. These bacteria have been called 'superbugs'.

The effect of antibiotics on infection

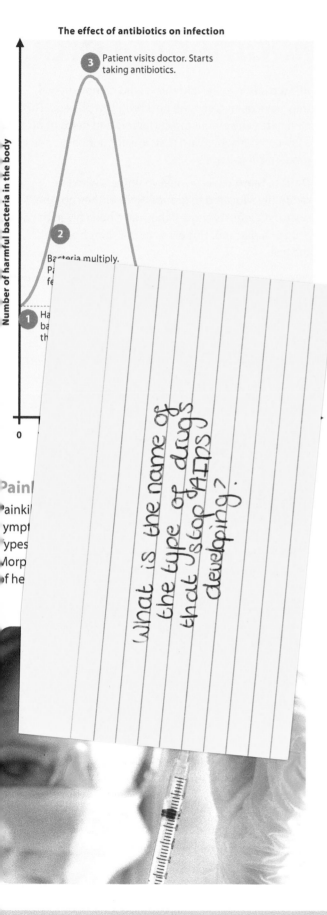

Number of harmful bacteria in the body

3 Patient visits doctor. Starts taking antibiotics.

2 Bacteria multiply.
P...
fe...

1 Ha...
ba...
th...

0

Painl
ainki
ympt
ypes
Morp
f he

(handwritten note): What is the name of the type of drugs that stop AIDS developing?

SUMMARY

- There are two types of vaccination: passive immunisation and active immunisation.

- Diseases caused by bacteria can be treated using antibiotics, but some bacteria are resistant to antibiotics.

- Painkillers do not kill pathogens; they only relieve symptoms.

QUESTIONS

QUICK TEST

1. What is an antibiotic?

2. Name the two types of vaccination.

3. What can doctors and patients do to reduce the risk of antibiotic-resistant bacteria developing?

EXAM PRACTICE

1. Explain how a vaccine works. **[4 marks]**

2. Bella is suffering from the flu and has asked her doctor for some antibiotics.

   a) How do antibiotics work? **[1 mark]**

   b) Explain why the doctor will not prescribe her antibiotics. **[2 marks]**

3. Explain why antibiotics are becoming increasingly less effective against 'superbugs' such as MRSA. **[3 marks]**

Discovery and development of drugs

Discovery of drugs

The following drugs are obtained from plants and microorganisms.

Name of drug	Use
Digitalis Foxgloves (common garden plants that are found in the wild)	Slows down the heartbeat; can be used to treat heart conditions
Aspirin From Willow trees (aspirin contains the active ingredient salicylic acid)	Mild painkiller
Penicillin Penicillium mould (discovered by Alexander Fleming)	Antibiotic

Modern pharmaceutical drugs are synthesised by chemists in laboratories, usually at great cost. The starting point might still be a chemical extracted from a plant.

New drugs have to be developed all the time to combat new and different diseases. This is a lengthy process, taking up to ten years. During this time the drugs are tested to determine:

- that they work
- that they are safe
- that they are given at the correct dose (early tests usually involve low doses).

In addition to testing, computer models are used to predict how the drug will affect cells, based on knowledge about how the body works and the effects of similar drugs. There are many who believe this type of testing should be extended and that animal testing should be phased out.

Clinical trials

Clinical trials are carried out on healthy volunteers and patients who have the disease. Some are given the new drug and others are given a placebo. The effects of the drug can then be compared to the effects of taking the placebo.

Blind trials involve volunteers who do not know if they have been given the new drug or a placebo. This eliminates any psychological factors and helps to provide a fair comparison. (Blind trials are not normally used in modern clinical trials.)

Double blind trials involve volunteers who are randomly allocated to groups. Neither they nor the doctors/scientists know if they have been given the new drug or a placebo. This eliminates all bias from the test.

New drugs must also be tested against the best existing treatments.

WS When studies involving new drugs are published, there is a **peer review**. This is where scientists with appropriate knowledge read the scientific study and examine the data to see if it is **valid**. Sometimes the trials are duplicated by others to see if similar results are obtained. This increases the **reliability** of the findings and filters out false or exaggerated claims.

Once a **consensus** is agreed, the paper is published. This allows others to hear about the work and to develop it further.

In the case of pharmaceutical drugs, clinical bodies have to decide if the drug can be **licensed** (allowed to be used) and whether it is **cost-effective**. This can be controversial because a potentially life-saving drug may not be used widely simply because it costs too much and/or would benefit too few people.

HT Monoclonal antibodies

Monoclonal antibodies are used to treat diseases and in technological applications. They are artificially produced in the laboratory.

The hybridoma cells are important because they divide rapidly (by cloning) and produce the required antibody. This means that millions of identical cloned cells are made, which can then be purified and used. See diagram opposite.

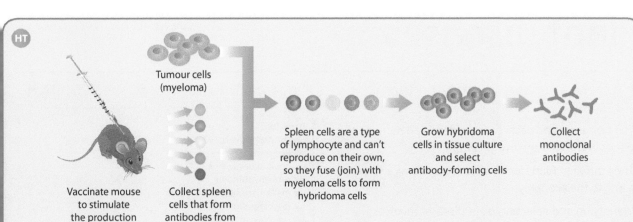

HT

Tumour cells (myeloma)

Vaccinate mouse to stimulate the production of antibodies

Collect spleen cells that form antibodies from mouse

Spleen cells are a type of lymphocyte and can't reproduce on their own, so they fuse (join) with myeloma cells to form hybridoma cells

Grow hybridoma cells in tissue culture and select antibody-forming cells

Collect monoclonal antibodies

Uses of monoclonal antibodies

Use	How?	Advantages
Cancer treatment	Antibodies are made that react to an antigen found on a cancer cell, e.g. in a pancreatic cancer tumour. A toxic agent, such as a radioactive substance, is bound to the antibody and can then be injected into the patient.	The antibody enables the agent to be carried directly to cancer cells and does not harm other cells in the body (unlike chemotherapy and radiotherapy). The technique can also be used to locate blood clots.
Pregnancy testing kits	HCG is a hormone produced in pregnancy. Antibodies to the protein hormone are bound to a test strip and used to detect HCG in urine.	Rapid results (within minutes).
Measuring hormone levels in laboratories	Antibodies can be used to detect hormones or to detect pathogens and other chemicals in the blood.	Provides new information that wouldn't otherwise have been available.
Research	Used to locate a specific molecule in cells or tissues by binding to them with a fluorescent marker.	Marking of molecules made more specific.

When using monoclonal antibodies to treat cancer, more side-effects than expected can arise.

SUMMARY

- Drugs used for treating illnesses and health conditions include antibiotics, analgesics and other chemicals that modify body processes and chemical reactions. In the past, these drugs came from plants and microorganisms.

- Clinical trials are carried out to ensure that drugs are safe to use.

- HT Monoclonal antibodies are a new technological advancement and have a variety of beneficial applications.

QUESTIONS

QUICK TEST

1. What is a double blind trial?
2. The drug, Digitalis, is obtained from foxglove plants. True or false?
3. HT List two uses of monoclonal antibodies.

EXAM PRACTICE

1. a) New pharmaceutical drugs have to be extensively tested before their release on to the market. Explain why this is. **[2 marks]**

   b) Name one alternative to animal testing of drugs. **[1 mark]**

   c) HT Explain how monoclonal antibodies can be used in pregnancy testing kits. **[2 marks]**

Plant diseases

HT Detecting and identifying plant diseases

Plant diseases and pests cause widespread damage to plants in the wild and in agriculture.

They threaten food stocks, affect ecosystems and cause a loss to the economy.

Identifying and detecting plant diseases involves several stages.

Process of detection and identification

Look for:
- stunted growth
- leaf spots
- areas of decay
- growths/tumours
- malformed stems and leaves
- discolouration
- presence of pests.

⬇

Eliminate environmental causes, e.g. abrasion, poor water supply, climatic factors.

⬇

Study the distribution of affected plants and refer to garden manuals or websites.

⬇

Carry out diagnostic testing in a laboratory to identify pathogens, e.g. observation; using monoclonal antibody testing kits; DNA testing.

Pathogens and pests that affect plants

Disease	Pathogen	Appearance/effect on plants		Treatment
Rose black spot	Fungal disease – the fungal spores are spread by water and wind	Purple/black spots on leaves; these then turn yellow and drop early, leading to a lack of photosynthesis and poor growth		Apply a fungicide and/or remove affected leaves
Tobacco mosaic virus (TMV)	Widespread disease that affects many plants (including tomatoes)	'Mosaic' pattern of discolouration; can lead to lack of photosynthesis and poor growth		
Ash dieback	Caused by the fungus Chalara	Leaf loss and bark lesions		
Barley powdery mildew	Erysiphe graminis	Causes powdery mildew to appear on grasses, including cereals		Fungicides and careful application of nitrogen fertilisers
Crown gall disease	Agrobacterium tumefaciens	Tumours or 'galls' at the crown of plants such as apple, raspberry and rose		Use of copper and methods of biological control

Pests	What they do	Appearance/effect on plants		Control
Invertebrates and particularly insects, e.g. many species of aphids	Feed on sap, leaves and storage organs; transmit pathogenic viruses			Chemical pesticides or biological control methods

Mineral ion deficiencies

Plants need **mineral ions** to build complex molecules. The ions are obtained from the soil via the roots in an active manner (requiring energy). In particular, plants need:

- **nitrates** to form **amino acids**, the building blocks of proteins. They are also needed to make nucleic acids such as DNA. Lack of nitrates in a plant leads to yellow leaves and stunted growth
- **magnesium** to form chlorophyll, which absorbs light energy for photosynthesis. Lack of magnesium results in chlorosis, which is a discolouration of the leaves.

Defence responses of plants

Plants defend themselves from herbivores and disease in a range of ways. These adaptations have evolved over millions of years to maximise a plant's survival.

Mechanical defences include:

- thorns and hairs to deter plant-eaters
- leaves that droop or curl on contact
- mimicry, to fool animals into avoiding them as food or laying their eggs on them. For example, passion flowers have structures that look like yellow eggs. A butterfly is less likely to lay eggs on a plant that has apparently already been used.

Physical defences include:

- tough, waxy leaf cuticles
- cellulose cell walls
- layers of dead cells around stems, e.g. bark. These fall off, taking pathogens with them.

Chemical defences include:

- producing antibacterial chemicals, e.g. mint and witch hazel
- producing toxins to deter herbivores, e.g. deadly nightshade, foxgloves and tobacco plants.

SUMMARY

- Pathogens can affect plants. Diseases include Rose black spot and Tobacco mosaic virus (TMV).
- Plants need minerals for healthy functioning, especially nitrates and magnesium.
- Plants defend themselves from pests and pathogens in three ways: mechanical defences, physical defences and chemical defences.

QUESTIONS

QUICK TEST

1. List three things that plant scientists look for when detecting whether a plant is diseased.

2. What type of disease is rose black spot?

3. What do plants need magnesium for?

EXAM PRACTICE

1. Black spot is a fungal disease that affects rose leaves.

   Fatima has noticed that roses growing in areas with high air pollution are less affected by the disease. She thinks that regular exposure to acid rain is stopping the disease from developing.

   Describe a simple experiment that Fatima could do to test her theory. **[3 marks]**

2. State one mechanical and one physical defence that plants use to defend themselves against herbivores or disease. **[2 marks]**

Homeostasis and negative feedback

Homeostasis

The body has automatic control systems to maintain a constant internal environment (**homeostasis**). These systems make sure that cells function efficiently.

Homeostasis balances inputs and outputs to ensure that optimal levels of temperature, pH, water, oxygen and carbon dioxide are maintained.

For example, even in the cold, homeostasis ensures that body temperature is regulated at about 37°C.

Control systems in the body may involve the nervous system, the endocrine system, or both. There are three components of control.

- **Effectors** cause responses that restore optimum levels, e.g. muscles and glands.

- **Coordination centres** receive and process information from the receptors, e.g. brain, spinal cord and pancreas.

- **Receptors** detect stimuli from the environment, e.g. taste buds, nasal receptors, the inner ear, touch receptors and receptors on retina cells.

WS When taking measurements, the quality of the measuring instrument and a scientist's skill is very important to achieve **accuracy**, **precision** and **minimal error**.

Adrenaline levels in blood plasma are measured by a chromatography method called HPLC. This is often coupled to a detector that gives a digital readout.

This digital readout displays the concentration of adrenaline as 6.32. This means that the instrument is precise up to $\frac{1}{100}$ of a unit.

A less precise instrument might only measure down to $\frac{1}{10}$ of a unit, e.g. 6.3 (one decimal place).

A bar graph of some data generated from HPLC is shown alongside. The graph shows the **mean concentration** of three different samples of blood. The average is taken from many individual measurements.

The vertical range bars indicate the range of measurements (the difference between highest and lowest) obtained for each sample.

Sample C shows **the greatest precision** as the individual readings do not vary as much as the others. There is less error so we can be more confident that the mean is closer to the **true value** and therefore more **accurate**.

HT Negative feedback

Negative feedback occurs frequently in homeostasis. It involves the automatic reversal of a change in the body's condition.

In the body, examples of negative feedback include osmoregulation/water balance, balancing blood sugar levels, maintaining a constant body temperature and controlling metabolic rate.

Metabolism needs to be controlled so that chemical reactions in the body take place at an optimal rate. Negative feedback controls metabolic rate by using the hormones **thyroxine** and **adrenaline**.

HT Thyroxine

Hypothalamus releases less TRH

Pituitary releases less TSH

Thyroid releases less thyroxine

Rise in metabolic rate

Controlling metabolic rate involves the interaction of the hormones **thyroxine**, thyroid-releasing hormone (**TRH**) and thyroid-stimulating hormone (**TSH**). Thyroxine stimulates the basal metabolic rate and plays an important role in growth and development.

Drop in metabolic rate

Thyroid releases more thyroxine

Pituitary releases more TSH

Hypothalamus releases more TRH

Adrenaline

Adrenaline is sometimes called the 'flight or fight' hormone. During times of stress the adrenal glands produce adrenaline. It has a direct effect on muscles, the liver, intestines and many other organs to prepare the body for sudden bursts of energy. Specifically, adrenaline increases the heart rate so that the brain and muscles receive oxygen and glucose more rapidly.

SUMMARY

- The body has control systems that work automatically to maintain a constant internal environment. This is homeostasis.
- Negative feedback is the automatic reversal of a change in the body's condition.

QUESTIONS

QUICK TEST

1. At what temperature is the human body regulated?
HT 2. Give two examples where negative feedback operates in the human body.

QUESTIONS

EXAM PRACTICE

HT 1. Match the organs below, with the function they perform:

Pancreas	Releases TSH
Skin receptor	Detects pressure
Pituitary gland	Releases TRH
Hypothalamus	Produces insulin

[3 marks]

The nervous system

Structure and function

The nervous system allows organisms to react to their surroundings and coordinate their behaviour.

The two main parts of the nervous system are:

- the central nervous system (**CNS**), which is made up of the spinal cord and brain
- the **peripheral nervous system**.

The flow of impulses in the nervous system is carried out by nerve cells linking the receptor, coordinator (neurones and synapses in the CNS) and effector.

The main components of the nervous system

CNS {
Brain
Spinal cord

The neurones that make up the peripheral nervous system

Sense organ	Sensory neurone	Synapse	Relay neurone	Synapse	Motor neurone	Muscle
In the sense organ, receptors detect a change – either inside or outside the body. The change is a stimulus.	Conducts the impulse from the sense organ towards the CNS.	The gap between the sensory and relay neurones.	Passes the impulse on to a motor neurone.	The gap between the relay neurone and the motor neurone.	Passes the impulse on to the muscle (or gland).	The muscle responds by contracting, which results in a movement. Muscles and glands are examples of effectors.

Nerve cells or neurones are specially adapted to carry nerve impulses, which are electrical in nature. The impulse is carried in the long, thin part of the cell called the axon.

Motor neurone

Fatty sheath

Axon

There are three types of neurone:

Sensory neurones carry impulses from receptors to the CNS.

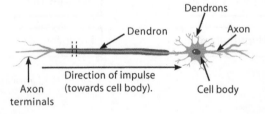

Dendrons

Dendron

Axon

Direction of impulse (towards cell body).

Axon terminals

Cell body

Relay neurones make connections between neurones inside the CNS.

Dendron

Axon terminals

Impulse travels first towards, and then away from, cell body.

Motor neurones carry impulses from the CNS to muscles and glands.

Muscle cells

Axon

Dendrons

Axon terminals

Direction of impulse (away from cell body).

Synapses

Synapses are junctions between neurones.

They play an important part in regulating the way impulses are transmitted. Synapses can be found between different neurones, neurones and muscles, and between dendrites (the root-like outgrowths from the cell body).

When an impulse reaches a synapse, a neurotransmitter is released by the neurone ('A' in the following diagram) into the gap that lies between the neurones. It travels by diffusion and binds to receptor molecules on the next neurone. This triggers a new electrical impulse to be generated.

Neurotransmitter released here — A synapse

Impulse

Neurone A

Receptor area — Cell body of neurone B

Electrical impulse generated in neurone B

WS You may have to investigate the effect of factors on human reaction time.

For example, you could be asked to investigate a learned reflex by measuring how far up a ruler someone can catch it. The nearer to the zero the ruler is caught, the faster the reflex.

You could investigate factors such as:

● experience/practice at catching

● sound ● touch ● sight.

Can you design experiments to test these variables? Which factors will need to be kept the same?

Reflex arcs

Reflex actions:

● are involuntary/automatic

● are very rapid

● protect the body from harm

● bypass conscious thought.

The pathway taken by impulses around the body is called a reflex arc. Examples include:

● opening and closing the pupil in the eye

● the knee-jerk response

● withdrawing your hand from a hot plate.

Here is the arc pathway for a pain response.

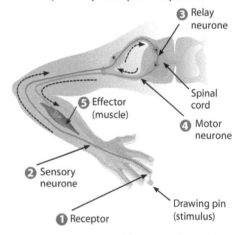

③ Relay neurone

Spinal cord

⑤ Effector (muscle)

④ Motor neurone

② Sensory neurone

Drawing pin (stimulus)

① Receptor

SUMMARY

● **The nervous system is made up of the brain and spinal cord (CNS) and neurones.**

● **There are three types of neurone: sensory, relay and motor neurones.**

● **Synapses are junctions between neurones where a neurotransmitter is released.**

● **A reflex arc is the pathway taken by impulses around the body in response to a reflex action.**

QUESTIONS

QUICK TEST

1. Name the junctions between neurones.

2. How do reflex arcs aid survival of an organism?

QUESTIONS

EXAM PRACTICE

1. The flow chart illustrates the events that occur during a reflex action.

   X → Receptor → Sensory neurone → Relay neurone → Brain and/or spinal cord → Motor neurone → Effector

   Paul accidentally puts his hand on a pin. Without thinking, he immediately pulls his hand away.

   a) Which component of a reflex arc is represented by the letter X? **[1 mark]**

   b) Give **two** reasons why this can be described as a reflex action. **[2 marks]**

   c) Use the features in the flow chart to describe what happens in this reflex action. **[4 marks]**

Nervous control

The brain

The brain is part of the central nervous system (**CNS**) and is studied by neuroscientists. The brain controls many activities in the body but is also responsible for memory, learning and behaviour.

The brain is made up of billions of neurones, each interlinked via synapses. The diagram shows the main structures in the brain.

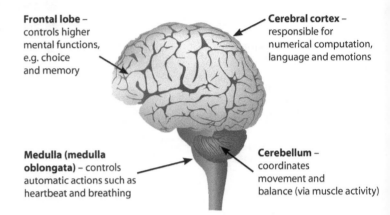

Frontal lobe – controls higher mental functions, e.g. choice and memory

Cerebral cortex – responsible for numerical computation, language and emotions

Medulla (medulla oblongata) – controls automatic actions such as heartbeat and breathing

Cerebellum – coordinates movement and balance (via muscle activity)

(HT) Studying the brain

Neuroscientists have mapped the regions of the brain to different functions. This is particularly useful for studying people suffering from brain damage and brain disorders, such as Alzheimer's disease. Studying healthy volunteers, using MRI technology and electrical stimulation of different parts of the brain, can reveal information about how this complex organ works.

Problems with studying and treating brain function include:

- obtaining subjects to study
- ethical issues relating to using human subjects
- difficulty when interpreting case studies
- difficulty in accessing areas of the brain where damage has occurred.

When treating nerve damage, the prospects can be limited because nervous and surrounding tissues are difficult to repair.

Scanning the brain

CT or **CAT** scans (computerised axial tomography) are X-ray tests producing cross-sectional images of the brain using computer programs. Many images can be taken at different angles to create 3D pictures.

PET stands for positron emission tomography. A PET scan can show what body tissues look like and how they are working. PET scanners work by detecting radiation given off by radiotracers, which the patient takes orally. The tracers accumulate in particular areas of the body. PET is often combined with CT and MRI technology to give 3D views of the brain.

Controlling body temperature

The thermoregulatory centre in the brain monitors and controls body temperature.

Enzymes work at around **37°C** (core temperature for humans). It is essential that this temperature does not fluctuate too much. Body heat is released by chemical reactions in the body (metabolism) and distributed via the bloodstream.

The skin plays a major role in temperature regulation. The structures responsible are found in the dermis and epidermis of the skin.

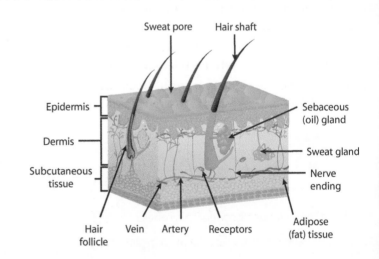

Sweat pore · Hair shaft

Epidermis

Dermis

Subcutaneous tissue

Sebaceous (oil) gland

Sweat gland

Nerve ending

Hair follicle · Vein · Artery · Receptors · Adipose (fat) tissue

When core temperature falls

- There is reduced blood flow near the skin to limit heat loss to the environment. This is called vasoconstriction.
- Tiny involuntary contractions of muscles (shivering) generate heat.
- Sweating reduces.
- Body hairs rise due to the contraction of special muscles. Humans have less body hair than most mammals so there is minimal benefit.

When core temperature rises

- Blood flows closer to the skin surface because blood vessels widen. This is called vasodilation.
- Sweating increases. The sweat evaporates on the skin's surface, which absorbs heat from the skin in an endothermic change.
- Body hairs lower.

Core temperature falling   **Core temperature rising**

Negative feedback system

There is a negative feedback system for regulating body temperature.

In this control system we have:

- receptors – thermoreceptors in the skin
- a controller – hypothalamus in the brain
- effectors – a variety of organs and tissues.

If a control system breaks down it can be life-threatening.

- **Hypothermia** is when the body temperature falls well below 37°C. Symptoms include shivering, blue skin colour, disorientation and unconsciousness.
- **Heat stroke/dehydration** is when the body temperature rises well above 37°C. Symptoms include an altered mental state, nausea, flushed skin, and rapid breathing and pulse.

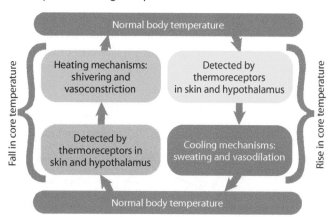

SUMMARY

- **The brain is part of the CNS and its main structures are the: frontal lobe, cerebral cortex, medulla and cerebellum.**
- **The thermoregulatory centre in the brain controls and monitors body temperature.**
- **The skin plays a major role in keeping core temperature within acceptable limits.**

QUESTIONS

QUICK TEST

1. What does CNS stand for?
2. What is vasodilation?

QUESTIONS

EXAM PRACTICE

HT 1. Jenny has been sunbathing on a beach where the temperature hasn't fallen below 35°C.

   Her body's thermoregulatory mechanisms have responded.

   Describe how negative feedback has operated to try to ensure that her core temperature does not rise much above 37°C.

   In your answer, explain how the thermoregulatory centre and the skin are involved. **[6 marks]**

The eye

Structure

The eye is a major sense organ in the human body. The receptors are on the **retina** – they consist of light-sensitive cells that detect the intensity and colour of light that enters the eye.

Light-sensitive cells are of two types – rods and cones. Rods are sensitive to low light intensities and enable vision in black and white. Cones are less sensitive and operate at higher light intensities. They enable us to see in colour. The three types of cone are those sensitive to:

- red light
- green light
- blue light.

Light entering the eye is perceived as a certain colour because these three types of cone are stimulated to different degrees, resulting in many millions of different shades and hues.

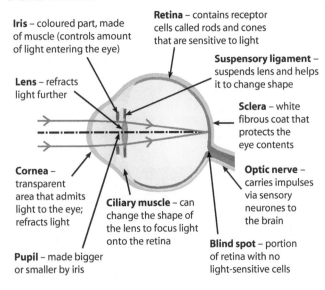

Iris – coloured part, made of muscle (controls amount of light entering the eye)

Retina – contains receptor cells called rods and cones that are sensitive to light

Suspensory ligament – suspends lens and helps it to change shape

Lens – refracts light further

Sclera – white fibrous coat that protects the eye contents

Cornea – transparent area that admits light to the eye; refracts light

Ciliary muscle – can change the shape of the lens to focus light onto the retina

Optic nerve – carries impulses via sensory neurones to the brain

Pupil – made bigger or smaller by iris

Blind spot – portion of retina with no light-sensitive cells

Accommodation

The function of the eye is to receive light rays and focus them to form a sharp image on the retina. The optical information is sent to the brain, which interprets it to give a sense of a 3D image, movement, colour and brightness. This interpretation by the occipital lobe in the brain is called **perception**.

The focusing process is called **accommodation**. It involves changing the shape of the lens in order to refract light.

To focus on a **near object**:

- light rays diverge significantly
- the ciliary muscles **contract**
- the suspensory ligaments loosen/go **slack**
- the lens becomes **short** and **fat** (**more convex**)
- the direction of light rays changes **greatly** to **converge** on the retina.

Ciliary muscle

Suspensory ligament

Lens

To focus on a **distant object**:

- light rays are almost parallel
- the ciliary muscles **relax**
- the suspensory ligaments become **taut**
- the lens becomes **long** and **thin** (**less convex**)
- the direction of light rays changes **slightly** to converge on the retina.

Ciliary muscle

Suspensory ligament

Lens

This is the front view of what happens during accommodation:

Near object

Suspensory ligaments slack

Ciliary muscles contracted

Lens more spherical

Distant object

Suspensory ligaments pulled taut

Ciliary muscles relaxed

Lens flat and thin

Defects

Common eye defects are:

- hyperopia (long sightedness)
- myopia (short sightedness)
- red–green colour blindness
- cataracts.

Hyperopia and myopia can be corrected using spectacles, contact lenses or laser surgery (to change the shape of the cornea).

Hyperopia is:

- caused by an eyeball that is too short, or a lens that stays too long and thin
- corrected by a **convex** lens.

Myopia is:

- caused by an eyeball that is too long, or weak suspensory ligaments that cannot pull the lens into a thin shape
- corrected by a **concave** lens.

Red–green colour blindness is more common in boys as it is determined by a defective gene on the X chromosome. The result is one or more cone types being faulty, causing an inability to distinguish between different shades of red or green.

A **cataract** develops when the lens clouds over. This is caused by some of the protein in the lens clumping together. Using spectacles with strong bifocals is a temporary treatment. The only permanent solution is removal during surgery.

An alternative to spectacles is **contact lenses**. These are miniature lenses (either hard or soft) that can be placed over the front of the eye. They need to be kept clean if they are a permanent type, or disposed of if temporary. Some operations can replace the natural lens with a new one.

WS You will be expected to give examples of how scientific ideas can lead to technical applications. The development of optical instruments for helping vision is a good example.

Sir Isaac Newton developed laws of optics in the seventeenth century. This built on theories put forward by scientists previously. An understanding of how light is refracted and reflected, and of ray diagrams, has led to inventions such as telescopes, microscopes and spectacles.

SUMMARY

- **The eye receives light rays and focuses them to form an image on the retina.**
- **The focusing process is called accommodation.**
- **Common eye defects are hyperopia and myopia.**

QUESTIONS

QUICK TEST

1. Which part of the eye contains rods and cones?

2. What causes myopia?

EXAM PRACTICE

1. Martin is reading a book when he hears a plane fly over above him. He shields his eyes against the sun and soon observes it as it flies into the distance.

   a) Explain the change that occurs in the pupil of Martin's eye as he observes the plane.

   State why this change needs to take place. **[3 marks]**

   b) In order to see the plane clearly, Martin's eyes need to focus on it.

   Describe how muscles in the eye affect the lenses in order to change the pathway of light rays. **[3 marks]**

The endocrine system

Structure and function

The endocrine system is made up of glands that are ductless and secrete **hormones** directly into the bloodstream. The blood carries these chemical messengers to **target organs** around the body, where they cause an effect.

Hormones:

- can be proteins or steroids
- interact with the nervous system to exert control over essential biological processes
- act over a longer time period than nervous responses but their effects are slower to establish.

Endocrine gland	Hormone(s) produced
Pituitary gland	TSH, ADH, FSH, LH, etc.
Pancreas	Insulin, glucagon
Thyroid	Thyroxine
Adrenal gland	Adrenaline
Ovaries (female)	Oestrogen, progesterone
Testes (male)	Testosterone

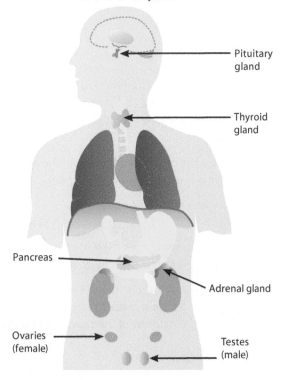

The endocrine system

Pituitary gland
Thyroid gland
Pancreas
Adrenal gland
Ovaries (female)
Testes (male)

The pituitary gland

The pituitary is often referred to as the **master gland** because it secretes many hormones that control other processes in the body. Pituitary hormones often trigger other hormones to be released.

Controlling blood glucose concentration

The control system for balancing blood glucose levels involves the **pancreas**.

The pancreas monitors the blood glucose concentration and releases hormones to restore the balance. When the concentration is too high, the pancreas produces insulin that causes glucose to be absorbed from the blood by all body cells, but particularly those in the liver and muscles. These organs convert glucose to glycogen for storage until required.

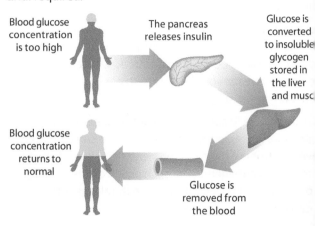

Blood glucose concentration is too high

The pancreas releases insulin

Glucose is converted to insoluble glycogen stored in the liver and musc

Blood glucose concentration returns to normal

Glucose is removed from the blood

HT If blood glucose concentration is too low, the pancreas secretes glucagon. This stimulates the conversion of glycogen to glucose via enzymic systems. It is then released into the blood.

Diabetes

There are two types of diabetes.

Type I diabetes:

- is caused by the pancreas' inability to produce insulin
- results in dangerously high levels of blood glucose
- is controlled by delivery of insulin into the bloodstream via injection or a 'patch' worn on the skin
- is more likely to occur in people under 40
- is the most common type of diabetes in childhood
- is thought to be triggered by an auto-immune response where cells in the pancreas are destroyed.

Type II diabetes:

- is caused by fatty deposits preventing body cells from absorbing insulin; the pancreas tries to compensate by producing more and more insulin until it is unable to produce any more, which results in dangerously high levels of blood glucose
- is controlled by a low carbohydrate diet and exercise initially; it may require insulin in the later stages
- is more common in people over 40
- is a risk if you are obese.

QUESTIONS

QUICK TEST

1. Which hormone is produced in the thyroid?
2. Where is insulin produced in the body?
3. What effects does type II diabetes have on the body?

EXAM PRACTICE

1. a) When someone eats a meal, the pancreas responds by producing insulin.

    Describe the effects this has on the body? **[2 marks]**

    b) People who have type I diabetes have trouble balancing their blood glucose concentrations.

    How is this condition treated? **[1 mark]**

    c) How is type II diabetes caused and managed? **[2 marks]**

Water and nitrogen balance

Excretion

Excretion is the process of getting rid of waste products made by chemical reactions in the body. Don't confuse it with **egestion**, which is the loss of solid waste (mainly undigested food).

The following are excreted products.

- Urea is made from the breakdown of excess amino acids in the liver. It is removed by the kidneys along with excess water and ions and transferred to the bladder as **urine** before being released.
- **Sweat** containing water, urea and salt is excreted by sweat glands onto the surface of the skin. Sweating aids the body's cooling process.
- **Carbon dioxide** and **water** are produced by respiration and leave the body from the lungs during exhalation.

The lungs and skin don't control the loss of substances. They are simply the organs by which these substances are removed.

The kidneys

Vena cava   Aorta
Kidney (filtration and reabsorption)
Renal artery } Blood supply to and from the kidneys
Renal vein
Bladder (stores urine)
Urethra (carries urine from bladder to outside)
Ureter (carries urine from kidney to bladder)

The kidneys filter the blood, allowing urea to pass to the bladder. The filtering is carried out within the kidney by thousands of tiny **kidney tubules**.

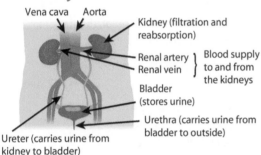

Kidney tubules
Capillary   Tubule
Water
Ions
Urea
Sugar
Water
Ions
Sugar
Most of the water, ions and all sugar   Excess water, ions and all urea

1. **Filtration**
Lots of water plus all the small molecules are squeezed out of the blood, under pressure, into the tubules.
2. **Selective reabsorption**
Useful substances, including glucose, ions and water, are reabsorbed into the blood from the tubules.
3. **Osmoregulation**
Amount of water in the blood and urine is adjusted here.
4. **Excretion of waste**
Excess water, ions and all the urea now pass to the bladder in the form of urine and are eventually released from the body.

Other useful substances (glucose, amino acids, fatty acids, glycerol and some water) are selectively re-absorbed early in the process.

The kidneys also control the balance of water in the blood. Damage occurs to red blood cells if water content is not balanced.

Ideal shape	Swollen	Shrivelled
When red blood cells (erythrocytes) are in solutions with equal concentration to their cytoplasm, they have an ideal, biconcave shape. This is because there is no net movement of water in or out.	When immersed in a solution of lower concentration (higher water concentration), the cells absorb water by osmosis. The weak cell membrane cannot resist the added water pressure and may burst.	In a more concentrated solution (lower water concentration), cells lose water by osmosis. They shrivel up and become **crenated** (have scalloped edges).

Kidney tubules (nephrons)

In terms of nephron structure, filtration, selective reabsorption and excretion of waste occur in the following regions.

Structure of the nephron

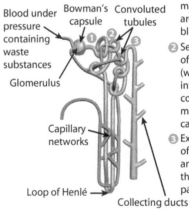

Blood under pressure containing waste substances
Bowman's capsule   Convoluted tubules
Glomerulus
Capillary networks
Loop of Henlé
Collecting ducts (lead to the ureter)

1. Filtration, where all small molecules and lots of water are squeezed out of the blood and into the tubules.
2. Selective reabsorption of useful substances (water, ions, glucose) back into the blood from the convoluted tubules. This may take energy in the case of glucose and ions.
3. Excretion of waste in the form of excess water, excess ions and all urea. These drain into the collecting tubules and pass to the bladder as urine.

Kidney failure

Kidneys may fail due to accidents or disease. A patient can survive with one kidney. If both kidneys are affected, two treatments are available.

Kidney transplant – involves a healthy person donating one kidney to replace two failed kidneys in another person.

Dialysis – offered to patients while they wait for the possibility of a kidney transplant. A dialysis machine removes urea and maintains levels of sodium and glucose in the blood.

his is what happens during dialysis.

1. Blood is taken from a person's vein and run into the dialysis machine, where it comes into close contact with a partially permeable membrane.

2. This separates the blood from the dialysis fluid.

3. The urea and other waste diffuse from the blood into the dialysis fluid. The useful substances remain and are transferred back to the body.

HT How urea is formed

Proteins obtained from the diet may produce a surplus of **amino acids** that need to be excreted safely.

1. First, the amino acids are deaminated in the **liver** to form ammonia.

2. Ammonia is toxic so is immediately converted to urea, which is then filtered out in the kidney.

Controlling water content

The osmotic balance of the body's fluids needs to be tightly controlled because if cells gain or lose too much water they do not function efficiently.

The amount of water re-absorbed by the kidneys is controlled by **anti-diuretic hormone (ADH)**. This is produced in the pituitary.

1. ADH directly increases the permeability of the kidney tubules to water.

2. When the water content of the blood is low (higher blood concentration), **negative feedback** operates to restore normal levels.

The effect of ADH on blood water content

Blood water level too low (salt concentration too high)	Blood water level too high (salt concentration too low)
Detected by the pituitary gland	Detected by the pituitary gland
More ADH released into the blood by pituitary gland	**Less** ADH released into the blood by pituitary gland
More water reabsorbed into the blood from the renal tubules	Less water reabsorbed into the blood from the renal tubules
Small amount of concentrated urine	Large amount of dilute urine
Normal blood water level	Normal blood water level

SUMMARY

- **Urea, sweat, carbon dioxide and water are the products of excretion.**

- **The kidneys remove urea from the body. When kidneys fail, a patient may receive dialysis or a kidney transplant.**

- **The kidney carries out water balance in tandem with the hormone ADH.**

QUESTIONS

QUICK TEST

1. List three substances that are selectively re-absorbed back into the bloodstream from the kidney.

2. Which organ releases ADH into the blood?

QUESTIONS

EXAM PRACTICE

1. A student is exercising.

   She produces carbon dioxide in her cells.

   a) What name is given to the process that produces carbon dioxide in the student's cells? **[1 mark]**

   b) Where in the body is carbon dioxide excreted? **[1 mark]**

   c) The student's liver produces urea.

      Which substances are changed to form urea? **[1 mark]**

Hormones in human reproduction

Puberty

During **puberty** (approximately 10–16 in girls and 12–17 in boys), the sex organs begin to produce **sex hormones**. This causes the development of secondary sexual characteristics.

In **males**, the primary sex hormone is **testosterone**.

During puberty, testosterone is produced from the testes and causes:

- production of sperm in testes
- development of muscles and penis
- deepening of the voice
- growth of pubic, facial and body hair.

In **females**, the primary sex hormone is **oestrogen**. Other sex hormones are **progesterone**, **FSH** and **LH**.

During puberty, oestrogen is produced in the ovaries and progesterone production starts when the menstrual cycle begins.

The secondary sexual characteristics are:

- ovulation and the menstrual cycle
- breast growth
- widening of hips
- growth of pubic and armpit hair.

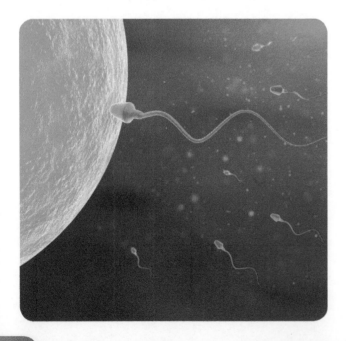

The menstrual cycle

A woman is fertile between the ages of approximately 13 and 50.

During this time, an egg is released from one of her ovaries each month and the lining of her uterus is replaced each month (approximately 28 days) to prepare for pregnancy.

The menstrual cycle

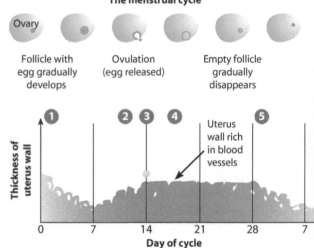

Ovary

Follicle with egg gradually develops

Ovulation (egg released)

Empty follicle gradually disappears

Uterus wall rich in blood vessels

Thickness of uterus wall

Day of cycle

0 7 14 21 28 7

Concentration of hormones in the blood

Oestrogen

Progesterone

1 Uterus lining breaks down (i.e. a period).

2 Repair of the uterus wall.
Oestrogen causes the uterus lining to gradually thicken.

3 Egg released by the ovary.

4 Progesterone and oestrogen make the lining stay thick, waiting for a fertilised egg.

5 No fertilised egg so cycle restarts.

As well as oestrogen and progesterone, the two other hormones involved in the cycle are:

- **FSH** or follicle stimulating hormone, which causes maturation of an egg in the ovary
- **LH** or luteinising hormone, which stimulates release of an egg.

HT Negative feedback in the menstrual cycle

The four female hormones interact in a complex manner to regulate the cycle.

- **FSH** is produced in the pituitary and acts on the ovaries, causing an egg to mature. It stimulates the ovaries to produce oestrogen.

- **Oestrogen** is secreted in the ovaries and inhibits further production of FSH. It also stimulates the release of LH and promotes repair of the uterus wall after menstruation.

- **LH** is produced in the pituitary. It also stimulates release of an egg.

- **Progesterone** is secreted by the empty follicle in the ovary (left by the egg). It maintains the lining of the uterus after ovulation has occurred. It also inhibits FSH and LH.

Female reproductive system

Low progesterone levels allow FSH from the pituitary gland to stimulate the maturation of an egg (in a follicle). This in turn stimulates oestrogen production.

High levels of oestrogen stimulate a surge in LH from the pituitary gland. This triggers ovulation in the middle of the cycle.

Brain

Pituitary gland

Uterus

Ovary

Ovary

SUMMARY

- During puberty the sex organs begin to produce sex hormones.
- Hormones play a vital role in regulating human reproduction, especially in the female menstrual cycle.

QUESTIONS

QUICK TEST

1. Name the four hormones involved in the menstrual cycle.

2. Name one effect of testosterone in puberty.

EXAM PRACTICE

1. **a)** State the functions of oestrogen and progesterone in the menstrual cycle. **[2 marks]**

   HT **b)** Oestrogen also stimulates release of LH.

   What is the result of this surge? **[1 mark]**

2. From the list below, circle **three** characteristics that develop as a result of puberty.

   Breast development   Fertilisation

   Sperm production   Secretion of ADH

   Menstrual cycle

   [3 marks]

Contraception and infertility

Non-hormonal contraception

Contraceptive method	Method of action	Advantages	Disadvantages
● Barrier method – condom (male + female)	Prevents the sperm from reaching the egg	● 82% effective ● Most effective against STIs	● Can only be used once ● May interrupt sexual activity ● Can break ● Some people are allergic to latex
● Barrier method – diaphragm	Prevents the sperm from reaching the egg	● 88% effective ● Can be put in place right before intercourse or 2–3 hours before ● Don't need to take out between acts of sexual intercourse	● Increases urinary tract infections ● Doesn't protect against STIs
● Intrauterine device	Prevents implantation – some release hormones	● 99% effective ● Very effective against pregnancy ● Doesn't need daily attention ● Comfortable ● Can be removed at any time	● Doesn't protect against STIs ● Needs to be inserted by a medical practitioner ● Higher risk of infection when first inserted ● Can have side effects such as menstrual cramping ● Can fall out and puncture the uterus (rare)
● Spermicidal agent	Kills or disables sperm	● 72% effective ● Cheap	● Doesn't protect against STIs ● Needs to be reapplied after one hour ● Increases urinary tract infections ● Some people are allergic to spermicidal agents
● Abstinence ● Calendar method	Refraining from sexual intercourse when an egg is likely to be in the oviduct	● 76% effective ● Natural ● Approved by many religions ● Woman gets to know her body and menstrual cycles	● Doesn't protect against STIs ● Calculating the ovulation period each month requires careful monitoring and instruction ● Can't have sexual intercourse for at least a week each month
● Surgical method	Vasectomy and female sterilisation	● 99% effective ● Very effective against pregnancy ● One-time decision providing permanent protection	● No protection against STIs ● Need to have minor surgery ● Permanent

Hormonal contraception

Contraceptive method	Method of action	Advantages	Disadvantages
● Oral contraceptive	Contains hormones that inhibit FSH production, so eggs fail to mature	● 91% effectiveness ● Very effective against pregnancy if used correctly ● Makes menstrual periods lighter and more regular ● Lowers risk of ovarian and uterine cancer, and other conditions ● Doesn't interrupt sexual activity	● Doesn't protect against STIs ● Need to remember to take it every day at the same time ● Can't be used by women with certain medical problems or by women taking certain medications ● Can occasionally cause side effects
● Hormone injection ● Skin patch ● Implant	Provides slow release of progesterone; this inhibits maturation and release of eggs	● 91–99% effectiveness depending on method used ● Lasts over many months or years ● Light or no menstrual periods ● Doesn't interrupt sexual activity	● Doesn't protect against STIs ● May require minor surgery (for implant) ● Can cause side effects

The percentage figures in the contraception tables are based on users in a whole population, regardless of whether they use the method correctly. If consistently used correctly, the percentage effectiveness of each method is usually higher. Some methods, such as the calendar method, are more prone to error than others.

HT Infertility treatment

Infertility treatment is used by couples who have problems conceiving.

Reasons for infertility

No eggs being released from the ovaries.
Endometriosis, which occurs when the tissue that lines the inside of the uterus enters other organs of the body, such as the abdomen and fallopian tubes; this reduces the maturation rate and release of eggs.
Male infertility/low sperm count.
Uterine fibroids.
Complete or partial blocking and/or scarring of the fallopian tubes.
Reduced number and quality of eggs.

Methods of treatment

Treating infertility is known as ART (assisted reproductive technology). There are a number of methods of treatment.

- Fertility drugs containing FSH and LH are given to women who do not produce enough FSH. They may then become pregnant naturally.
- Clomifene therapy can help women who struggle to become pregnant because they do not ovulate. Clomifene is a drug that can stimulate ovulation.

In vitro fertilisation (IVF) is a method in which the potential mother is given FSH and LH to stimulate the production of several eggs. Sperm is collected from the father. The sperm and eggs are then introduced together outside the body in a petri dish. One or two growing embryos can then be transplanted into the woman's uterus.

IVF under a microscope

Injected with father's sperm

Holds the egg in place

Unfertilised human egg cell

Disadvantages of IVF

It is very expensive.
It can be mentally and physically stressful.
Success rates are only approximately 40% (at the time of writing).
There is an increased risk of multiple births.

SUMMARY

- **Fertility and the possibility of pregnancy can be controlled using non-hormonal and hormonal methods of contraception.**
- **Infertility has a number of causes and may be treated using fertility drugs and surgical procedures.**

QUESTIONS

QUICK TEST

1. Which non-hormonal contraceptive is the least effective? What is the benefit of this method?

2. Why are condoms effective against the spread of HIV?

EXAM PRACTICE

1. State one advantage and one disadvantage of using a condom barrier method of contraception. **[2 marks]**

HT 2. IVF is one method of treating infertility.

   Explain how this is carried out. **[3 marks]**

Plant hormones

Tropisms and general control

Plants, as well as animals, respond to changes in their environment.

Plant hormones are chemicals that control the growth of shoots and roots, flowering and the ripening of fruits.

Two major groups of hormones are:

- auxins
- gibberellins.

These hormones move through the plant in solution and affect its growth by responding to:

- gravity (gravitropism/geotropism)
- light (phototropism).

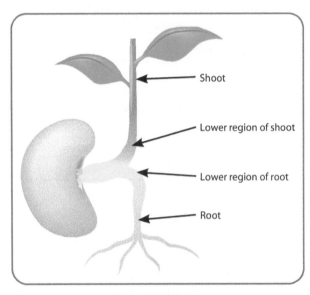

- Shoot
- Lower region of shoot
- Lower region of root
- Root

Response to gravity

- **Shoots** grow against gravity (negative gravitropism).
- **Roots** grow in the direction of gravity (positive gravitropism).

In the roots, auxins inhibit growth in the lower region, which makes the roots grow downwards. The roots anchor the plant in the soil and seek out water and minerals for absorption.

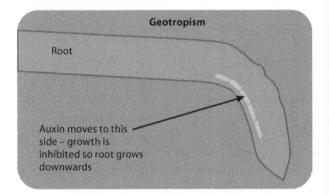

Geotropism

Root

Auxin moves to this side – growth is inhibited so root grows downwards

Response to light

Roots grow away from the light (negative phototropism)

Shoots grow towards the light (positive phototropism).

In the shoot, auxin is made in the tip. When light reaches the tip from all directions, the hormone is equally distributed. When light is **uni-directional** (in this case from the right), the auxin moves to the shaded side, causing the cells here to **elongate**. This makes the shoot bend towards the light.

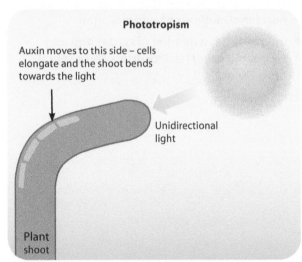

Phototropism

Auxin moves to this side – cells elongate and the shoot bends towards the light

Unidirectional light

Plant shoot

HT Commercial uses of hormones

Auxins

● **Rooting powder** consists of an auxin that encourages the growth of roots in stem cuttings, so that many new plants can be grown from a parent.

● **Tissue culture** technique also requires auxins to promote growth.

● **Selective weed killers**, used in agriculture, contain auxins that disrupt the growth patterns of their target plants (which are often broad-leaved rather than narrow-leaved). Therefore the crop plant is not harmed.

Gibberellins

Gibberellins are important in initiating seed germination. They can be used to:

● end seed dormancy
● promote flowering
● increase fruit size
● produce seedless fruit (parthenocarpy).

Ethene

Ethene is a hydrocarbon that controls cell division. It can be used in the food industry to control ripening of fruit while it is being transported or stored.

WS One of your required practicals is likely to be investigating the effect of light on the growth of newly germinated shoots.

This is one possible investigation. What other investigations could you carry out?

Experiment to show that shoots grow towards light

1 Cut a hole in the side of a box. Put three cuttings into the box. The cuttings detect light coming from the hole and will grow towards it.

2 Cut a hole in the side of another box. Put three cuttings with foil-covered tips in the box. These shoots can't detect the light so they grow straight up.

SUMMARY

● Plant hormones control the growth of shoots and roots, flowering, and ripening of fruits.

● There are two major groups of plant hormones: auxins and gibberellins.

● The hormones respond to gravity and light.

HT ● Commercial uses of plant hormones include plant propagation, weed killers and improved fruit production.

QUESTIONS

QUICK TEST

1. What name is given to the process where shoots grow towards the light?

HT 2. Which hormone could a farmer apply to her crop to kill weeds?

HT 3. Which hormone could be used to cause plants to flower early?

EXAM PRACTICE

1. Which substance does rooting powder contain that stimulates root growth?

   Tick (✓) one box.   [1 mark]

   Enzymes   ☐

   Nitrates   ☐

   Plant hormones   ☐

   Vitamins   ☐

2. The diagram shows a bean seedling that has been grown horizontally in laboratory conditions.

   Explain what has happened to cause the growth in the shoot at point 'A' in the diagram.   [2 marks]

Sexual and asexual reproduction

Sexual reproduction

Sexual reproduction is where a male **gamete** (e.g. sperm) meets a female gamete (e.g. egg). This fusion of the two gametes is called fertilisation and may be internal or external.

Gametes are produced by meiosis in the sex organs.

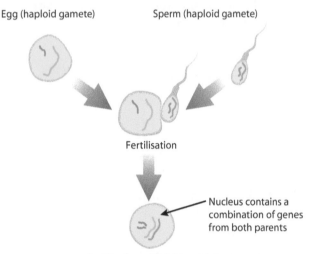

Egg (haploid gamete)   Sperm (haploid gamete)

Fertilisation

Nucleus contains a combination of genes from both parents

Fertilised egg (diploid zygote)

Asexual reproduction

Asexual reproduction does not require different male and female cells. Instead, genetically identical clones are produced from mitosis. These may just be individual cells, as in the case of yeast, or whole multicellular organisms, e.g. aphids.

Many organisms can reproduce using both methods, depending on the environmental conditions.

A yeast budding

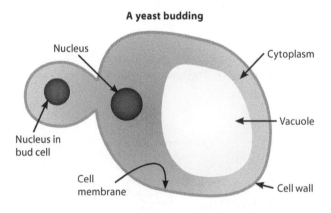

Nucleus

Cytoplasm

Vacuole

Nucleus in bud cell

Cell membrane

Cell wall

Comparing sexual and asexual reproduction

Advantages of sexual reproduction

- Produces **variation** in offspring through the process of meiosis, where genes are 'shuffled'. Variation is increased by **random fusion of gametes**.
- Survival advantage gained when the environment changes because different genetic types have more chance of producing well-adapted offspring.
- Humans can make use of sexual reproduction through **selective breeding**. This enhances food production.

Disadvantages of sexual reproduction

- Relatively slow process.
- Variation can be a disadvantage in stable environments.
- More resources required than for asexual reproduction, e.g. energy, time.
- Results of selective breeding are unpredictable and might lead to genetic abnormalities from 'in-breeding'.

Advantages of asexual reproduction

- Only one parent required.
- Fewer resources (energy and time) need to be devoted to finding a mate.
- Faster than sexual reproduction – survival advantage of producing many offspring in a short period of time.
- Many identical offspring of a well-adapted individual can be produced to take advantage of favourable conditions.

Disadvantages of asexual reproduction

- Offspring may not be well adapted in a changing environment.

Different reproductive strategies

Malarial parasite

The **plasmodium** is a **protist** that causes malaria. It can reproduce asexually in the human host but sexually in the mosquito.

Parasite enters human from mosquito bite

Parasite re-enters mosquito when it feeds

Strawberry plants and daffodils

These plants can reproduce sexually using flowers, or asexually using **runners** (strawberry) or **bulb division** (daffodils).

Strawberry plant

Runner

Daffodils

Bulb

Fungi

Many fungi, such as toadstools, mushrooms and moulds, can reproduce sexually (giving variation) or asexually by **spores**.

WS During your course, you will learn about historical developments in science and technology. It is important to appreciate how understanding develops from previous studies and their publication in scientific journals. Scientists build on each other's work and applications develop from this.

Malaria is a disease that kills many people every year. Treatment via anti-malarial drugs has been available for years and techniques are available to disrupt the parasite's life cycle. At the time of writing, scientists are close to developing a vaccine.

All these developments have come from an understanding of the complex interaction between the plasmodium and mosquito's life cycles. Studies were carried out by many different people and took about ten years to reach their conclusions. Finally, a Scottish physician called Sir Ronald Ross produced evidence for the complete life cycle of the malaria parasite in mosquitoes. For this work, he received the 1902 Nobel Prize in Medicine.

SUMMARY

- One of the basic characteristics of life is reproduction. This is the means by which a species continues. If sufficient offspring are not produced, the species becomes extinct.
- Reproduction may be sexual or asexual.
- Some organisms reproduce using both means.

QUESTIONS

QUICK TEST

1. What name is given to the fusion of two gametes?

2. How do strawberry plants reproduce asexually?

3. How do fungi reproduce asexually?

EXAM PRACTICE

1. State one advantage and one disadvantage of reproducing by sexual means. **[2 marks]**

2. a) The malarial parasite can live in two different organisms.

      Suggest a survival advantage in this. **[1 mark]**

   b) Name one other type of organism that combines asexual and sexual reproduction. **[1 mark]**

DNA

DNA and the genome

The nucleus of each cell contains a complete set of genetic instructions called the **genetic code**. The information is carried as genes, which are small sections of DNA found on **chromosomes**. The genetic code controls cell activity and, consequently, characteristics of the whole organism.

DNA facts

- DNA is a polymer.
- It is made of two strands coiled around each other called a **double helix**.
- The genetic code is in the form of nitrogenous **bases**.
- Bases bond together in pairs forming hydrogen bond cross-links.
- The structure of DNA was discovered in 1953 by **James Watson** and **Francis Crick**, using experimental data from **Rosalind Franklin** and **Maurice Wilkins**.
- A single gene codes for a particular sequence of **amino acids**, which, in turn, make up a single **protein**.

The human genome

The **genome** of an organism is the entire genetic material present in an adult cell of an organism.

The Human Genome Project (HGP)

The HGP was an international study. Its purpose was to map the complete set of genes in the human body.

HGP scientists worked out the code of the human genome in three ways. They:

- determined the sequence of all the bases in the genome
- drew up maps showing the locations of the genes on chromosomes

- produced linkage maps that could be used for tracking inherited traits from generation to generation, e.g. for genetic diseases. This could then lead to targeted treatments for these conditions.

The results of the project, which involved collaboration between UK and US scientists, were published in 2003. Three billion base pairs were determined.

The mapping of the human genome has enabled anthropologists to work out historical human migration patterns. This has been achieved by collecting and analysing DNA samples from many people across the globe. The study is called the **Genographic Project**.

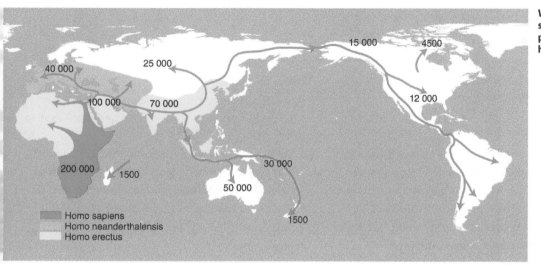

World map showing suggested migration patterns of early hominids

15 000
4500
25 000
40 000
100 000
70 000
12 000
200 000
1500
30 000
50 000
1500

Homo sapiens
Homo neanderthalensis
Homo erectus

DNA structure and base sequences

The four bases in DNA are A, T, C and G. The code is 'read' on one strand of DNA. Three consecutive bases (a **triplet**) code for one particular amino acid. The sequence of these triplets determines the structure of a whole protein.

The bases are attached to a sugar phosphate **backbone**. These form a basic unit called a nucleotide.

A section of DNA

T — A
G — C
C — G
A — T

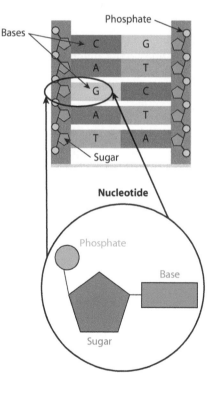

Bases
Phosphate
C G
A T
G C
A T
T A
Sugar

Nucleotide

Phosphate
Base
Sugar

SUMMARY

● Each cell in the body contains a nucleus with chromosomes.

● Genes are small sections of DNA found on chromosomes. The four bases in DNA are A, T, C and G.

● The Human Genome Project has produced a complete map of all the genes in the body.

QUESTIONS

QUICK TEST

1. Explain how the Human Genome Project has advanced medical science.

2. What are the four bases in DNA?

EXAM PRACTICE

1. Explain how scientists have used knowledge about the human genome to contribute to the **Genographic project**. [2 marks]

2. Describe how a nucleotide from the DNA molecule is made up. [3 marks]

The genetic code

HT DNA structure and base sequences

In complementary strands of DNA, T always bonds with A, and C with G.

Proteins are synthesised in structures called **ribosomes**, which are located in the cytoplasm of the cell. In order for the DNA code in the nucleus to be translated as new protein by the ribosomes, the following process occurs.

1 DNA unzips and exposes the bases on each strand.

2 A molecule of **messenger RNA (mRNA)** is constructed from one of these template strands.

3 The mRNA, which carries a complementary version of the gene, travels out of the nucleus to the ribosome in cytoplasm.

4 In the ribosome, the mRNA is 'read'. **Transfer RNA (tRNA)** molecules carry individual amino acids to add to a growing protein (polypeptide).

5 The new polypeptide folds into a unique shape and is released into the cytoplasm.

In the nucleus, transcription takes place.

In the ribosome, translation takes place.

The proteins that are produced carry out a specific function. Examples include enzymes, hormones or structural protein such as **collagen**.

Gene switching

Every cell contains a complete set of genes for the whole body, but only some of these genes are used in any one cell.

- Some genes are not **expressed**. They are said to be 'switched off'. Switching off genes is accomplished through **non-coding** DNA. Most DNA is non-coding.
- The genes that are 'switched on' eventually determine the function of a cell.

Mutations

Mutations (genetic variants):

- are changes to the structure of a DNA molecule
- occur continuously during the cell division process or as a result of external influences, e.g. exposure to radioactive materials or emissions such as X-rays or UV light
- usually have a neutral effect as the same amino acids may be coded for or the proteins produced work in the same way
- may result in harmful conditions or, more rarely, beneficial traits
- result in a change in base sequence and therefore changes in the amino acid sequence and protein structure.

The G base is substituted for an A base

Proteins produced as a result of mutation may no longer be able to carry out their function. This is because they have a different 3D structure. For example, an enzyme's active site may no longer fit with its substrate, or a structural protein may lose its strength.

Changes in the base sequence may be passed on to daughter cells when cell division occurs. This in turn may lead to offspring having genetic conditions.

Monohybrid crosses

Most characteristics or **traits** are the result of multiple genes interacting but some are controlled by a single gene. Examples include fur colour in mice and the shape of ear lobes in humans.

These genes exist as a pair of alleles on homologous chromosomes.

Alleles are described as **dominant** or **recessive**.

- A dominant allele controls the development of a characteristic even if it is present on only one chromosome in a pair.
- A recessive allele controls the development of a characteristic only if a dominant allele is not present, i.e. if the recessive allele is present on both chromosomes in a pair.

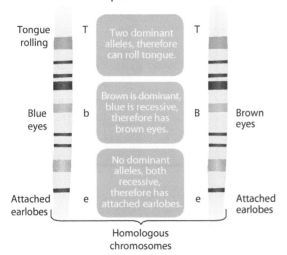

Homologous chromosomes

If both chromosomes in a pair contain the same allele of a gene, the individual is described as being **homozygous** for that gene or condition.

If the chromosomes in a pair contain different alleles of a gene, the individual is **heterozygous** for that gene or condition.

The combination of alleles for a particular characteristic is called the **genotype**. For example, the genotype for a homozygous dominant tongue-roller would be TT. The fact that this individual is able to roll their tongue is termed their **phenotype**.

Other examples are:

- bb (genotype), blue eyes (phenotype)
- EE or Ee (genotype), unattached/pendulous ear lobes (phenotype).

When a characteristic is determined by just one pair of alleles, as with eye colour and tongue rolling, it is called **monohybrid inheritance**.

WS During your course, you will be expected to recognise, draw and interpret scientific diagrams.

In this module, the way complementary strands of DNA are arranged can be worked out once you know that base T bonds with A and base C bonds with G.

Can you write out the complementary (DNA) strand to this sequence?

| A | T | T | A | C | G | T | G | A | G | C | C |

SUMMARY

- **HT** **Bases in DNA always bond T to A and G to C.**
- **HT** **Mutations are genetic variants and may result in harmful conditions or occasionally beneficial traits.**
- Alleles can be dominant or recessive.

QUESTIONS

QUICK TEST

1. Name the two different types of alleles.
2. **HT** What are mRNA and tRNA?

EXAM PRACTICE

1. Explain how a recessive allele could control the development of a characteristic. **[2 marks]**

2. **HT** Compare and contrast the locations and roles of transfer RNA and messenger RNA. **[4 marks]**

DAY 6 / ⏱ 45 Minutes

Inheritance and genetic disorders

Genetic diagrams

Genetic diagrams are used to show all the possible combinations of alleles and outcomes for a particular gene. They use:

- capital letters for dominant alleles
- lower-case letters for recessive alleles.

For eye colour, brown is dominant and blue is recessive. So B represents a brown allele and b represents a blue allele.

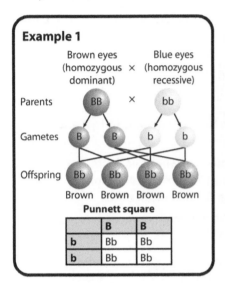

Example 1

Punnett square

	B	B
b	Bb	Bb
b	Bb	Bb

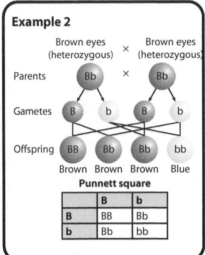

Example 2

Punnett square

	B	b
B	BB	Bb
b	Bb	bb

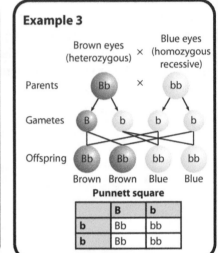

Example 3

Punnett square

	B	b
b	Bb	bb
b	Bb	bb

(WS) You need to be able to interpret genetic diagrams and work out ratios of offspring.

- In Example 2 above, the phenotypes of the offspring are 'brown eyes' and 'blue eyes'. As there are potentially three times as many brown-eyed children as blue-eyed, the phenotypes are said to be in a 3:1 ratio.

- In Example 3 above, the ratio would be 1:1 because half of the theoretical offspring are brown-eyed and half blue-eyed. Another way of saying this is that the probability of parents producing a brown-eyed child is 50%, or ½, or 0.5.

There might be a number of different alleles for each gene, rather than just two. This is called multiple alleles. E.g. in the inheritance of blood groups there are three possible alleles in the **ABO** system, but each person can only have two of these alleles.

(HT) In exams, you may be asked to construct your own Punnett squares to solve genetic cross problems.

Family trees

Family trees are another way of showing how genetic traits can be passed on. Here is an example.

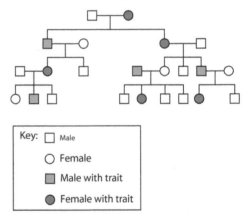

Key:
- ☐ Male
- ○ Female
- ◼ Male with trait
- ● Female with trait

Inheritance of sex

Sex in humans/mammals is determined by whole chromosomes. These are the 23rd pair and are called sex chromosomes. There is an 'X' chromosome and a smaller 'Y' chromosome. The other 22 chromosome pairs carry the remainder of genes coding for the rest of the body's characteristics.

All egg cells carry X chromosomes. Half the sperm carry X chromosomes and half carry Y chromosomes. The sex of an individual depends on whether the egg is fertilised by an X-carrying sperm or a Y-carrying sperm.

If an X sperm fertilises the egg it will become a girl. If a Y sperm fertilises the egg it will become a boy. The chances of these events are equal, which results in approximately equal numbers of male and female offspring.

Inherited diseases

Some disorders are caused by a 'faulty' gene, which means they can be **inherited**. One example is **polydactyly**, which is caused by a dominant allele and results in extra fingers or toes. The condition is not life-threatening.

Cystic fibrosis, on the other hand, can limit life expectancy. It causes the mucus in respiratory passages and the gut lining to be very thick, leading to build-up of phlegm and difficulty in producing correct digestive enzymes.

Cystic fibrosis is caused by a recessive allele. This means that an individual will only exhibit symptoms if both recessive alleles are present in the genotype. Those carrying just one allele will not show symptoms, but could potentially pass the condition on to offspring. Such people are called **carriers**.

Conditions such as cystic fibrosis are mostly caused by **faulty alleles** that are **recessive**.

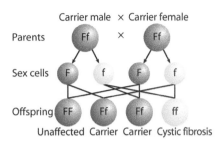

Knowing that there is a 1 in 4 chance that their child might have cystic fibrosis gives parents the opportunity to make decisions about whether to take the risk and have a child. This is a very difficult decision to make.

Technology has advanced and it is now possible to screen embryos for genetic disorders.

● If an embryo has a life-threatening condition it could be destroyed, or simply not be implanted if IVF was applied.

● Alternatively, new gene therapy techniques might be able to reverse the effects of the condition, resulting in a healthy baby.

As yet, these possibilities have to gain approval from ethics committees – some people think that this type of 'interference with nature' could have harmful consequences.

SUMMARY

● **Genetic diagrams show all possible combinations of alleles and outcomes for a gene. Family trees are another way of showing how genetic traits are passed on.**

QUESTIONS

QUICK TEST

1. If a homozygous brown-eyed individual is crossed with a homozygous blue-eyed individual, what is the probability of them producing a blue-eyed child?

EXAM PRACTICE

1. A man who knows he is a carrier of cystic fibrosis is thinking of having children with his partner. His partner does not suffer from cystic fibrosis, nor is she a carrier.

   Suggest what a genetic counsellor might advise the couple.

   Use genetic diagrams to aid your explanation and state any probabilities of offspring produced. **[6 marks]**

Variation and evolution

Variation

The two major factors that contribute to the appearance and function of an organism are:

- **genetic information** passed on from parents to offspring

- **environment** – the conditions that affect that organism during its lifetime, e.g. climate, diet, etc.

These two factors account for the large variation we see **within** and **between** species. In most cases, both of these factors play a part.

Evolution

Put simply, evolution is the theory that all organisms have arisen from simpler life forms over billions of years. It is driven by the **mechanism** of **natural selection**. For natural selection to occur, there must be genetic variation between individuals of the same species. This is caused by mutation or new combinations of alleles resulting from sexual reproduction.

Most mutations have no effect on the phenotype of an organism. Where they do, and if the environment is changing, this can lead to relatively rapid change.

Evidence for evolution

Evidence for evolution comes from many sources. It includes:

- comparing genomes of different organisms

- studying embryos and their similarities

- looking at changes in species during modern times, e.g. antibiotic resistance in bacteria

- comparing the anatomy of different forms

- the fossil record.

One of the earliest sources of evidence for evolution was the discovery of fossils.

Further evidence for evolution – the pentadactyl limb

If you compare the forelimbs of a variety of vertebrates, you can see that the bone structures are all variations on a five-digit form, whether it be a leg, a wing or a flipper. This suggests that there was a common ancestral form from which these organisms developed.

Human   Cow   Horse   Whale   Bird

How fossils are formed

1 When an animal or plant dies, the processes of decay usually cause all the body tissues to break down. In rare circumstances, the organism's body is rapidly covered and oxygen is prevented from reaching it. Instead of decay, fossilisation occurs.

2 Over hundreds of thousands of years, further sediments are laid down and compress the organism's remains.

3 Parts of the organism, such as bones and teeth, are **replaced by minerals** from solutions in the rock.

4 Earth upheavals, e.g. **tectonic plate movement**, bring sediments containing the fossils nearer the Earth's surface.

5 Erosion of the rock by wind, rain and rivers exposes the fossil. At this stage, the remains might be found and excavated by **paleontologists**.

Fossils can also be formed from footprints, burrows and traces of tree roots.

By comparing different fossils and where they are found in the rock layers, paleontologists can gain insights into how one form may have developed into another.

Difficulties occur with earlier life forms because many were **soft-bodied** and therefore not as well-preserved as organisms with bones or shells. Any that are formed are easily destroyed by Earth movements. As a result of this, scientists cannot be certain about exactly how life began.

Extinction

The fossil record provides evidence that most organisms that once existed have become **extinct**. In fact, there have been at least five **mass** extinctions in geological history where most organisms died out. One of these coincides with the disappearance of dinosaurs.

Causes of extinction include:

- **catastrophic events**, e.g. volcanic eruptions, asteroid collisions
- changes to the environment over geological time
- new **predators**
- new **diseases**
- new, more successful **competitors**.

QUESTIONS

QUICK TEST

1. Suggest two pieces of evidence that support the theory of evolution through natural selection.
2. Suggest two causes of extinction.

EXAM PRACTICE

1. The fore-limbs found in human, cow, horse, whale and bird are all based on the pentadactyl limb which means a five-digited limb.

   Scientists believe that the whale may have evolved from a horse-like ancestor that lived in swampy regions millions of years ago.

   Suggest how whales could have evolved from a horse-like mammal. In your answer, use Darwin's theory of natural selection. **[5 marks]**

SUMMARY

- **Genetic information, and environment are the two major factors that contribute to variation.**
- **Evolution is the theory that all organisms have developed from simple life forms over billions of years.**
- **The fossil record tells us that most organisms that once existed have become extinct.**
- **Speciation occurs where members of an original population can no longer interbreed.**

Darwin and evolution

Human evolution

Modern man has evolved from a common ancestor that gave rise to all the primates: gorillas, chimpanzees and orangutans. DNA comparisons have shown we are most closely related to the chimpanzee.

The evolution of man can be traced back over the last four to five million years. Over this period of time, the **human form** (hominid) has developed:

● a more upright, bipedal stance

● less body hair

● a smaller and less 'domed' forehead

● greater intelligence and use of tools, initially from stone. These tools can be dated using scientific techniques, e.g. radiometric dating and carbon dating.

There have been significant finds of fossils that give clues to human evolution.

① **Ardi** – at 4.4 million years old, this is the oldest, most complete hominid skeleton.

② **Lucy** – from 3.2 million years ago, this is one of the first fossils to show an upright walking stance.

③ **Proconsul skull** – discovered by **Mary Leakey** and her husband; the hominid is thought to be from about 1.6 million years ago.

Darwin's theory of evolution through natural selection

Within a population of organisms there is a range of variation among individuals. This is caused by genes. Some differences will be beneficial; some will not.

Beneficial characteristics make an organism more likely to survive and pass on their genes to the next generation. This is especially true if the environment is changing. This ability to be successful is called **survival of the fittest**.

Species that are not well adapted to their environment may become extinct. This process of change is summed up in the theory of evolution through **natural selection**, put forward by **Charles Darwin** in the nineteenth century.

Many theories have tried to explain how life might have come about in its present form.

However, Darwin's theory is accepted by most scientists today. This is because it explains a wide range of observations and has been discussed and tested by many scientists.

Darwin's theory can be reduced to five ideas.

They are:
● variation ● competition ● survival of the fittest
● inheritance ● extinction.

Darwin's ideas are illustrated in the following two examples.

Example 1: peppered moths

Variation – most peppered moths are pale and speckled. They are easily camouflaged amongst the lichens on silver birch tree bark. There are some rare, dark-coloured varieties (that originally arose from genetic mutation). They are easily seen and eaten by birds.

Competition – in areas with high levels of air pollution, lichens die and the bark becomes discoloured by soot. The lighter peppered moths are now put at a competitive disadvantage.

Survival of the fittest – the dark (melanic) moths are now more likely to avoid detection by pedators.

Inheritance – the genes for dark colour are passed on to offspring and gradually become more common in the general population.

Extinction – if the environment remains polluted, the lighter form is more likely to become extinct.

Dark peppered moth

Peppered moth

Example 2: methicillin-resistant bacteria

The resistance of some bacteria to antibiotics is an increasing problem. MRSA bacteria have become more common in hospital wards and are difficult to eradicate.

Variation – bacteria mutate by chance, giving them a resistance to antibiotics.

Competition – the non-resistant bacteria are more likely to be killed by the antibiotic and become less competitive.

Survival of the fittest – the antibiotic-resistant bacteria survive and reproduce more often.

Inheritance – resistant bacteria pass on their genes to a new generation; the gene becomes more common in the general population.

Extinction – non-resistant bacteria are replaced by the newer, resistant strain.

To slow down the rate at which new, resistant strains of bacteria can develop:

- doctors are urged not to prescribe antibiotics for obvious viral infections or for mild bacterial infections

- patients should complete the full course of antibiotics to ensure that **all** bacteria are destroyed (so that none will survive to mutate into resistant strains).

Species become more and more specialised as they evolve and adapt to their environmental conditions.

The point at which a new species is formed occurs when the original population can no longer interbreed with the newer, 'mutant' population. For this to occur, **isolation** needs to happen.

Speciation

- Groups of the same species that are separated from each other by physical boundaries (like mountains or seas) will not be able to breed and share their genes. This is called **geographical isolation**.

- Over long periods of time, separate groups may specialise so much that they cannot successfully breed any longer and so two new species are formed – this is **reproductive isolation**.

SUMMARY

- **The ability to be successful is called survival of the fittest. Species that are not well adapted to their environment may become extinct.**
- **This process is the basis of the theory of evolution by natural selection, put forward by Charles Darwin.**

QUESTIONS

QUICK TEST

1. Define evolution.

2. What is the name of the oldest most complete hominid?

3. State two examples where natural selection has been observed by scientists in recent times.

EXAM PRACTICE

1. Alfred Wallace lived at the same time as Charles Darwin and also wrote papers on the idea of natural selection. In 1858 he wrote the following:

"I ask myself: how and why do species change, and why do they change into new and well-defined species. Why and how do they become so exactly adapted to distinct modes of life; and why do all the intermediate grades die out?"

Wallace had no knowledge of genes or DNA structure.

Knowing what we know now, how would you answer his question about how organisms become 'exactly adapted?' **[2 marks]**

Evolution – the modern synthesis

The pioneers of evolution

Charles Darwin published his findings and theories in a book called *On the Origin of Species* (1859). At the time, the reaction to Darwin's theory, particularly from religious authorities, was hostile. They felt he was saying that 'people were descended from monkeys' (although he wasn't) and that he denied God's role in the creation of man. This meant that Darwin's theory was only slowly and reluctantly accepted by many people in spite of his many eminent supporters. In Darwin's time there wasn't the wealth of evidence we have today from the study of genetics and DNA. (The structure of DNA wasn't made plain until 1953.)

Jean-Baptiste Lamarck had published ideas about the gradual change in organisms before Darwin but he put forward a different approach (now known to be inaccurate). He believed that evolution was driven by the inheritance of acquired characteristics. However, he thought that:

● organisms changed during their lifetime as they struggled to survive

● these changes were passed on to the offspring.

For example, he said that when giraffes stretched their necks to reach leaves higher up, this extra neck length was passed on to their offspring.

Lamarck's theory was rejected because there was no evidence that the changes occurring in an individual's lifetime could alter their genes and so be passed on to their offspring. The difference between Darwin and Lamarck's approach is shown in the picture.

Lamarck believed that the necks of giraffes stretched during their lifetime to reach food in trees. They then passed this characteristic on to the next generation.

Darwin believed giraffes that had longer necks could reach more food in trees, so they were more likely to survive and reproduce (survival of the fittest).

Alfred Russel Wallace also put forward a theory of evolution by natural selection. In fact, he and Darwin jointly published writings a year before Darwin's *On the Origin of Species*.

Wallace gathered evidence from around the world to back up the theory of natural selection. In particular, he looked at how speciation could occur and wrote extensively about warning colouration in animals.

Gregor Mendel carried out breeding programmes for plants in the middle of the nineteenth century. He discovered that characteristics could be passed down as 'units' (which we now know as genes) in a predictable way.

> **HT** **WS** The significance of Mendel's work was only realised after his death. This shows the importance of a scientist's work being published so that others can build on it. It also shows that, sometimes, apparently trivial phenomena and their study can lead to important theories. In this way, our understanding of the world we live in is enhanced.
>
> In your course, you will need to give examples of **how a model can be tested** by **observation** and **experiment**. Mendel's work is a classic example of this because he could show that the ratios of offspring in his pea plant experiments could be accurately predicted.

Other evidence to support the theory of natural selection

Since the nineteenth century, a sequence of findings and techniques has further reinforced our ideas about evolution by natural selection.

● In the late nineteenth century, chromosomes were observed during cell division under the microscope.

● In the early twentieth century, the link between Mendel's 'units' and genes was made. The idea that genes were found on chromosomes was put forward.

● In the mid-twentieth century, the structure of DNA was worked out by Watson and Crick.

Cloning

Studies of genetics have led to many technological advancements. One of these is the ability to clone cells, organs or even whole organisms.

Cuttings

When a gardener has a plant with all the desired characteristics, he/she may choose to produce lots of them by taking stem, leaf or root cuttings. The cuttings are grown in a damp atmosphere until roots develop.

Cloning by tissue culture

To mass-produce plants that are genetically identical to the parent plant and to each other, horticulturalists follow this method.

1 Select a parent plant with the characteristics that you want.
2 Scrape off lots of small pieces of tissue into several beakers containing nutrients and hormones.
3 A week or two later there will be lots of genetically identical plantlets growing.
4 Repeat the process.

Embryo transplantation

Embryo transplantation is now commonly used to breed farm animals.

1 Sperm is collected from an adult male with desirable characteristics.
2 A selected female is artificially inseminated with the male animal's sperm.
3 The fertilised egg develops into an embryo that is removed from the mother at an early stage.
4 In the laboratory, the embryo is split to form several clones.
5 Each clone is transplanted into a female who will be the surrogate mother to the newborn.

Adult cell cloning

1 The diploid nucleus is taken from a mature cell (ordinary body cell) of the donor organism.
2 The diploid nucleus, containing all of the donor's genetic information, is inserted into an empty egg cell (i.e. an egg cell with the nucleus removed or enucleated). This is nuclear transfer. A tiny electric shock is applied at this stage to trigger cell division in the embryo.
3 The egg cell, containing the diploid nucleus, is stimulated so that it begins to divide by mitosis.
4 The resulting embryo is implanted in the uterus of a 'surrogate mother'.
5 The embryo develops into a foetus and is born as normal.

Diploid nucleus from a body cell containing genetic information

Empty (enucleated) egg cell

Surrogate mother

Clone (Dolly)

QUESTIONS

EXAM PRACTICE

1. Describe the main sequence involved in mass producing a desirable plant through tissue culture. The first step has been done for you: [3 marks]

    – Select parent with desired characters ...

2. Dolly the sheep was the first animal to be produced by adult cell cloning, but she only lived until the age of 8 years and it took over 100 attempts to produce a viable embryo. Some say, this is too high a price to pay for this technique.

    State the advantages of adult cell cloning to counter this argument.   [2 marks]

QUESTIONS

QUICK TEST

1. How did Darwin's mechanism of natural selection differ from Lamarck's?

2. Describe the advantage of cloning plants or animals.

Selective breeding and genetic engineering

Selective breeding

Farmers and dog breeders have used the principles of selective breeding for thousands of years by keeping the best animals and plants for breeding.

For example, to breed Dalmatian dogs, the spottiest dogs have been bred through the generations to eventually get Dalmatians. The factor most affected by selective breeding in dogs is probably temperament. Most breeds are either naturally obedient to humans or are trained to be so.

This is the process of selective breeding.

This process has to be repeated many times to get the desired results.

Advantages of selective breeding
● It results in an organism with the 'right' characteristics for a particular function.
● In farming and horticulture, it is a more efficient and economically viable process than natural selection.

Disadvantages of selective breeding
● Intensive selective breeding reduces the gene pool – the range of alleles in the population decreases so there is **less variation**.
● Lower variation reduces a species' ability to respond to environmental change.
● It can lead to an accumulation of harmful recessive characteristics (in-breeding), e.g. restriction of respiratory pathways and dislocatable joints in bulldogs.

Examples of selective breeding

Modern food plants

Three of our modern vegetables have come from a single ancestor by selective breeding. (Remember, it can take many, many generations to get the desired results.)

Selective breeding in plants has also been undertaken to produce:

● disease resistance in crops

● large, unusual flowers in garden plants.

Modern cattle

Selective breeding can contribute to improved yields in cattle. Here are some examples.

● **Quantity of milk** – years of selecting and breeding cattle that produce larger than average quantities of milk has produced herds of cows that produce high daily volumes of milk.

● **Quality of milk** – as a result of selective breeding, Jersey cows produce milk that is rich and creamy, and can therefore be sold at a higher price.

● **Beef production** – the characteristics of the Hereford and Angus varieties have been selected for beef production over the past 200 years or more. They include hardiness, early maturity, high numbers of offspring and the swift, efficient conversion of grass into body mass (meat).

Genetic engineering

All living organisms use the same basic genetic code (DNA). So genes can be transferred from one organism to another in order to deliberately change the recipient's characteristics. This process is called genetic engineering or genetic modification (GM).

Altering the genetic make-up of an organism can be done for many reasons.

● **To improve resistance to herbicides**: for example, soya plants are genetically modified by inserting a gene that makes them resistant to a herbicide. When the crop fields are sprayed with the herbicide only the weeds die, leaving the soya plants without competition so they can grow better. Resistance to frost or disease can also be genetically engineered. Bigger yields result.

● **To improve the quality of food**: for example, bigger and more tasty fruit.

● **To produce a substance you require**: for example, the gene for human insulin can be inserted into bacteria or fungi, to make human insulin on a large scale to treat diabetes.

● **Disease resistance**: crop plants receive genes from the bacterium *Bacillus thuringiensis*, which allows them to produce insecticide in their cells.

Advantages of genetic engineering
● It allows organisms with new features to be produced rapidly.
● It can be used to make biochemical processes cheaper and more efficient.
● In the future, it may be possible to use genetic engineering to change a person's genes and cure certain disorders, e.g. cystic fibrosis. This is an area of research called gene therapy.

Disadvantages of genetic engineering
● Transplanted genes may have unexpected harmful effects on human health.
● Some people are worried that GM plants may cross-breed with wild plants and release their new genes into the environment.

HT Producing insulin

The following method is used to produce insulin.

1. The human gene for insulin production is identified and removed using a restriction enzyme, which cuts through the DNA strands in precise places.

2. The same restriction enzyme is used to cut open a ring of bacterial vector DNA (a plasmid).

3. Other enzymes called ligases are then used to insert the section of human DNA into the plasmid. The DNA can be 'spliced' in this way because the ends of the strands are 'sticky'.

4. The plasmid is reinserted into a bacterium, which starts to divide rapidly. As it divides, it replicates the plasmid.

5. The bacteria are cultivated on a large scale in fermenters. Each bacterium carries the instructions to make insulin. When the bacteria make the protein, commercial quantities of insulin are produced.

Part of a human DNA strand

Desired gene

Desired gene isolated

Desired gene inserted into target DNA

Target bacterium replicates itself

Bacterium

Sometimes, other vectors are used to introduce human DNA into organisms, e.g. viruses. It is important that the hybrid genes are transferred to the host organism at an early stage of its development, e.g. the egg or the embryo stage. As the cells are quite undifferentiated, the desired characteristics from the inserted DNA are more likely to develop.

SUMMARY

● Selective breeding is done to ensure desirable characteristics are passed on.

● Genetic modification is when genes are transferred from one organism to another to change the recipient's characteristics.

QUESTIONS

QUICK TEST

1. What advantages does genetic engineering have over selective breeding?

HT 2. What is a plasmid?

EXAM PRACTICE

1. State one advantage and one disadvantage of selective breeding. **[2 marks]**

HT 2. Describe the process where artificial insulin is produced via genetic engineering.

State any enzymes involved and how the final product is obtained. The first stage is completed for you. **[5 marks]**

Stage 1: Human gene for insulin identified

Stage 2: ....

Stage 3: ....

Stage 4: ....

Stage 5: ....

Stage 6: ....

Classification

The origins of classification

In the past, observable characteristics were used to place organisms into categories.

In the eighteenth century, **Carl Linnaeus** produced the first classification system. He developed a hierarchical arrangement where larger groups were subdivided into smaller ones.

Kingdom	Largest group
Phylum	
Class	
Order	
Family	
Genus	
Species	Smallest group

Linnaeus also developed a binomial system for naming organisms according to their genus and species. For example, the common domestic cat is *Felis catus*. Its full classification would be:

- Kingdom: *Animalia*
- Phylum: *Chordata*
- Class: *Mammalia*
- Order: *Carnivora*
- Family: *Felidae*
- Genus: *Felis*
- Species: *Catus*.

Linnaeus' system was built on and resulted in a **five-kingdom system**. Developments that contributed to the introduction of this system included improvements in microscopes and a more thorough understanding of the biochemical processes that occur in all living things. For example, the presence of particular chemical pathways in a range of organisms indicated that they probably had a common ancestor and so were more closely related than organisms that didn't share these pathways.

Kingdom	Features	Examples
Plants	Cellulose cell wall Use light energy to produce food	Flowering plants Trees Ferns Mosses
Animals	Multicellular Feed on other organisms	Vertebrates Invertebrates
Fungi	Cell wall of chitin Produce spores	Toadstools Mushrooms Yeasts Moulds
Protoctista Protozoa	Mostly single-celled organisms	Amoeba Paramecium
Prokaryotes	No nucleus	Bacteria Blue–green algae

The classification diagram below illustrates how different lines of evidence can be used. The classes of vertebrates share a common ancestor and so are quite closely related. Evidence for this lies in comparative anatomy (e.g. the pentadactyl limb) and similarities in biochemical pathways.

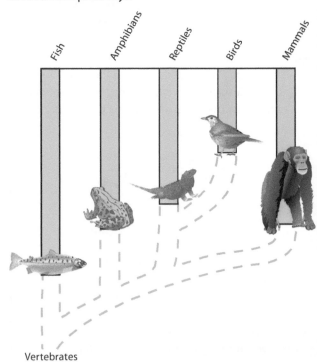

Vertebrates

In more recent times, improvements in science have led to a **three-domain system** developed by **Carl Woese**. In this system organisms are split into:

- **archaea** (primitive bacteria)

- **bacteria** (true bacteria)
- **eukaryota** (including Protista, fungi, plants and animals).

Further improvements in science include chemical analysis and further refinements in comparisons between non-coding sections of DNA.

Evolutionary trees

Tree diagrams are useful for depicting relationships between similar groups of organisms and determining how they may have developed from common ancestors. Fossil evidence can be invaluable in establishing these relationships.

Here, two species are shown to have evolved from a common ancestor.

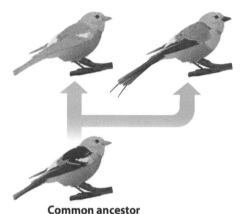

Common ancestor

WS You need to understand how new evidence and data leads to changes in models and theories.

In the case of the three-domain system, a more accurate and cohesive classification structure was proposed as a result of improvements in microscopy and increasing knowledge of organisms' internal structures.

These apes share a common ancestor

SUMMARY

- There is a huge variety of living organisms. Scientists group or classify them using shared characteristics.
- This is important because it helps to: work out how organisms evolved on Earth; understand how organisms coexist in ecological communities and identify and monitor rare organisms that are at risk from extinction.

QUESTIONS

QUICK TEST

1. What do the first and second words in the binomial name of an organism mean?

2. Who developed the recent three-domain system?

EXAM PRACTICE

1. In the Linnean system of classification, organisms are placed within a hierarchy of organisation.

   a) The following table shows this hierarchy with some missing levels. Complete the missing parts.

Kingdom
....
Class
....
Family
....
Species

   [3 marks]

   b) Linnaeus also developed the bionomial system of classification. The domestic dog has the following classification:

   Kingdom – *Animalia*; Class – *Mammalia*; Family – *Canidae*; Genus – *Canis*; Species – *familiaris*

   Use this information to suggest the binomial name of the domestic dog. **[1 mark]**

Organisms and ecosystems

Communities

Ecosystems are physical environments with a particular set of conditions (**abiotic** factors), plus all the organisms that live in them. The organisms interact through competition and predation. An ecosystem can support itself without any influx of other factors or materials. Its energy source (usually the Sun) is the only external factor.

Other terms help to describe aspects of the environment.

- The **habitat** of an animal or plant is the part of the physical environment where it lives. There are many types of habitat, each with particular characteristics, e.g. pond, hedgerow, coral reef.
- A **population** is the number of individuals of a species in a defined area.
- A **community** is the total number of individuals of all the different populations of organisms that live together in a habitat at any one time.

An organism must be well-suited to its habitat to be able to compete with other species for limited environmental resources. Even organisms within the same species may compete in order to survive and breed. Organisms that are specialised in this way are restricted to that type of habitat because their adaptations are unsuitable elsewhere.

Resources that plants compete over include:
- light
- space
- water
- minerals.

Animals compete over:
- food
- mates.
- territory.

Interdependence

In communities, each species may depend on other species for food, shelter, pollination and seed dispersal. If one species is removed, it may have knock-on effects for other species.

Stable communities contain species whose numbers fluctuate very little over time. The populations are in balance with the physical factors that exist in that habitat. Stable communities include **tropical rainforests** and ancient oak **woodlands**.

Adaptations

Adaptations:

- are special features or behaviours that make an organism particularly well-suited to its environment and better able to compete with other organisms for limited resources
- can be thought of as a biological solution to an environmental challenge – evolution provides the solution and makes species fit their environment.

Animals have developed in many different ways to become well adapted to their environment and to help them survive. Adaptations are usually of three types:

- **Structural** – for example, skin colouration in chameleons provides camouflage to hide them from predators.
- **Functional** – for example, some worms have blood with a high affinity for oxygen; this helps them to survive in anaerobic environments.
- **Behavioural** – for example, penguins huddle together to conserve body heat in the Antarctic habitat.

Look at the **polar bear** and its life in a very cold climate. It has:

- small ears and large bulk to reduce its surface area to volume ratio and so reduce heat loss
- a large amount of insulating fat (blubber)
- thick white fur for insulation and camouflage
- large feet to spread its weight on snow and ice
- fur on the soles of its paws for insulation and grip
- powerful legs so it is a good swimmer and runner, which enables it to catch its food
- sharp claws and teeth to capture prey.

The **cactus** is well adapted to living in a desert habitat. It:

- has a rounded shape, which gives a small surface area to volume ratio and therefore reduces water loss
- has a thick waxy cuticle to reduce water loss
- stores water in a spongy layer inside its stem to resist drought
- has sunken stomata, meaning that air movement is reduced, minimising loss of water vapour through them
- has leaves that take the form of spines to reduce water loss and to protect the cactus from predators.

Some organisms have biochemical adaptations. **Extremophiles** can survive extreme environmental conditions. For example:

- bacteria living in deep sea vents have optimum temperatures for enzymes that are much higher than 37°C
- icefish have antifreeze chemicals in their bodies, which lower the freezing point of body fluids
- some organisms can resist high salt concentrations or pressure.

(ws) During your course you will be asked to suggest explanations for observations made in the field or laboratory. These include:

- suggesting factors for which organisms are competing in a certain habitat
- giving possible adaptations for organisms in a habitat.

For example, low-lying plants in forest ecosystems often have specific adaptations for maximising light absorption as they are shaded by taller plants. Adaptations might include leaves with a large surface area and higher concentrations of photosynthetic pigments to absorb the correct wavelengths and lower intensities of light.

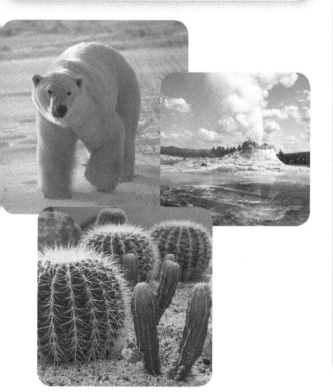

SUMMARY

- In a community, each species may depend on other species for food, shelter, pollination and seed dispersal. They are interdependent.
- Organisms need to be well adapted within the ecosystem in order to survive.
- Adaptations may be structural, functional or behavioural.
- Extremophiles can survive in extreme environments, e.g. very low temperatures.

QUESTIONS

QUICK TEST

1. What is an ecosystem?

2. What three resources do animals compete over?

3. Give two examples of extremophiles.

EXAM PRACTICE

1. Amazonian tree frogs have adaptations suited to living in equatorial rainforests.

   Explain how each of the adaptations listed below helps the frog to survive.

   i) Sticky pads on its 'fingers' and toes.

   ii) Hides its bright colours when asleep by closing its eyes and tucking its feet under its body.

   iii) Long sticky tongue. **[3 marks]**

2. What is an extremophile?

   Give one example of how these organisms are well adapted to their habitats. **[2 marks]**

Studying ecosystems 1

Taking measurements in ecosystems

Ecosystems involve the interaction between **non-living** (**abiotic**) and **living** (**biotic**) parts of the environment. So it is important to identify which factors need to be measured in a particular habitat.

Abiotic factors include:
- light intensity
- temperature
- moisture levels
- soil pH and mineral content
- wind intensity and direction
- carbon dioxide levels for plants
- oxygen levels for aquatic animals.

Biotic factors include:
- availability of food
- new predators arriving
- new pathogens
- one species out-competing another.

Measuring biotic factors – sampling methods

It is usually impossible to count all the species living in a particular area, so a sample is taken.

When sampling, make sure you:
- **take a big enough sample** to make the estimate good and reliable – the larger the sample, the more accurate the results.
- **sample randomly** – the more random the sample, the more likely it is to be representative of the population.

Quadrats

Quadrats are square frames that typically have sides of length 0.5 m. They provide excellent results as long as they are placed randomly. The population of a certain species can then be estimated.

For example, if an average of 4 dandelion plants are found in a 0.25 m^2 quadrat, a scientist would estimate that 16 dandelion plants would be found in each 1 m^2 and 16 000 dandelion plants in a 1000 m^2 field.

A quadrat   0.5 m
Area = 0.25 m^2 0.5 m

Transects

Sometimes an environmental scientist may want to look at how species change across a habitat, or the boundary

between two different habitats – for example, the plants found in a field as you move away from a hedgerow.

This needs a different approach that is systematic rather than random.

1. Lay down a line such as a tape measure. Mark regular intervals on it.
2. Next to the line, lay down a small quadrat. Estimate or count the number of plants of the different species. This can sometimes be done by estimating the percentage cover.
3. Move the quadrat along at regular intervals. Estimate and record the plant populations at each point until the end of the line.

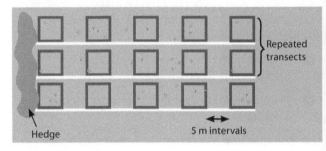

Repeated transects

Hedge   5 m intervals

Sampling methods

Sampling animal populations is more problematic as they are mobile and well adapted to evade capture. Here are four of the main techniques used.

Pooters
Insects sucked in here · You suck here · Fine mesh to stop you from sucking the insects into your mouth	This is a simple technique in which insects are gathered up easily without harm. With this method, you get to find out which species are actually present, although you have to be systematic about your sampling in order to get representative results and it is difficult to get ideas of numbers.

Sweepnets
	Sweepnets are used in long grass or moderately dense woodland where there are lots of shrubs. Again, it is difficult to get truly representative samples, particularly in terms of the relative numbers of organisms.

Pitfall traps
	Pitfall traps are set into the ground and used to catch small insects, e.g. beetles. Sometimes a mixture of ethanol or detergent and water is placed in the bottom of the trap to kill the samples, and prevent them from escaping. This method can give an indication of the relative numbers of organisms in a given area if enough traps are used to give a representative sample.

Capture/recapture

The capture/recapture method is sometimes called the Lincoln index.

1. Animals are caught humanely – for example, woodlice are caught in traps overnight. Their number is counted and recorded.

2. The animals are marked in some way – for example, water boatmen (a type of insect) can be marked with a drop of waterproof paint on their upper surface.

3. The marked animals are then released back into the population for a suitable amount of time.

4. A second sample is obtained, which will contain some marked animals and some unmarked. The numbers in each group are again counted.

Certain assumptions are made when using capture/recapture data. These include:

- no death, immigration or emigration
- each sample being collected in exactly the same way without bias
- the marks given to the animals not affecting their survival rate, e.g. using paint on invertebrates requires care because if too much is added it can enter their respiratory passages and kill them.

The following formula can then be used to estimate the total population size in the habitat.

$$\text{population size} = \frac{\text{number in first sample (all marked)} \times \text{number in second sample (marked and unmarked)}}{\text{number in second sample that were previously marked}}$$

Example

Fifty-six mice are caught in woodland and a small section of fur removed from their tails with clippers. The mice are released. The next evening, a further sample of sixty-two mice are caught. Twenty-five of these mice have shaved tails. What is the total population size?

Population size $= \dfrac{56 \times 62}{25} = 139$ mice

Keys

Correctly identifying species in a sample can be difficult. Using keys like this one can help to identify organisms.

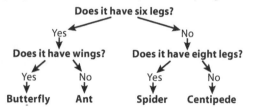

SUMMARY

- Ecosystems involve interaction between abiotic and biotic parts of the environment.
- Biotic factors can be measured by sampling methods such as quadrats and transects.
- Animal numbers can be sampled using trapping techniques.

QUESTIONS

QUICK TEST

1. What information does a transect give you?

2. What is the difference between biotic and abiotic factors?

EXAM PRACTICE

1. A group of students were carrying out a quadrat survey to determine the population of daisies in a park. The park measured 60 m × 90 m.

   Their results are shown in the table.

Quadrat	1	2	3	4	5	6	7	8	9	10
No. daisies	5	2	1	0	4	5	2	0	6	3

   a) They placed the quadrats randomly.

   Why was this?   **[1 mark]**

   b) Calculate the median number of daisies per quadrat.   **[3 marks]**

   c) Another student calculated the mean number of daisies per quadrat to be 2.3.

   If the quadrats measured 1 m × 1 m, calculate the estimated number of daisies in the park.

   Show your working.   **[2 marks]**

Studying ecosystems 2

Measuring abiotic factors

Many measurements connected with climate and the weather can be measured using meteorological equipment. Measuring chemical pollutant levels is particularly useful as they give information about concentrations of pollutants. This can be done by **direct chemical tests**.

For example, water can be tested for pH and samples can be assessed for metal ion content, e.g. mercury.

Indicator species

The occurrence of certain indicator species can be observed by sampling an area and noting whether the species is present or not, together with the overall numbers of the species.

For example, insect larvae and other invertebrates can act as an indicator of water pollution. When sewage works outflow into a stream, this pollutes the water by altering the levels of nitrogen compounds in the stream and reducing oxygen levels. This has an impact on the organisms that can survive in the stream.

Organisms that can cope with pollution include the rat-tailed maggot, the bloodworm, the water louse and the sludge worm. However, some organisms are very sensitive to this type of water pollution so they are not found in areas where the levels are high.

Organisms such as mayfly and stonefly larvae are killed by high levels of water pollution (they cannot tolerate low oxygen levels), so they are indicators of clean water.

rat-tailed maggot   stonefly nymph

Air quality can be measured using lichens as **indicator species**.

Lichens ➡

Few lichens indicate high concentration of sulfur dioxide in the air

Lichens ➡

Many lichens indicate low concentration of sulfur dioxide in the air

Predator–prey relationships

Animals that kill and eat other animals are called **predators** (e.g. foxes, lynx). The animals that are eaten are called **prey** (e.g. rabbits, snowshoe hares).

Many animals can be both predator and prey. For instance, a stoat is a predator when it hunts rabbits and it is the prey when it is hunted by a fox.

Predator – stoat

Prey – rabbit

Predator – fox

Prey – stoat

In nature there is a delicate balance between the population of a predator (e.g. lynx) and its prey (e.g. snowshoe hare). However, the prey will always outnumber the predators.

The number of predators and prey follow a classic population cycle. There will always be more hares than lynx and the population peak for the lynx will always come after the population peak for the hare. As the population cycle is cause and effect, they will always be out of phase.

SUMMARY

- Indicator species can be observed by sampling an area to see whether the species is present or not.
- Predators kill and eat prey. The prey will always outnumber predators. Numbers of predators and prey vary in cycles.

QUESTIONS

QUICK TEST

1. What is an indicator species?

2. What is a predator?

EXAM PRACTICE

1. Jenna has been using a net to capture organisms from a fast-flowing stream. After half an hour, she counts and identifies the organisms she has found.

   Her results are displayed in the table.

Species	Number
Stonefly nymphs	1
Mayfly nymphs	5
Caddis fly larvae	2
Water fleas	75

   a) Jenna says that because of the indicator species she has found, the water is fairly unpolluted.

      Do you agree?

      Give reasons for your answer. **[2 marks]**

   b) Apart from further samples, what other data would she need to collect in order to verify this conclusion? **[1 mark]**

2. Rabbits and stoat numbers tend to follow a cyclical pattern as their numbers change over the years.

   Explain why this is so. **[3 marks]**

Recycling

The water cycle

Water is a vital part of the biosphere. Most organisms consist of over 50% water.

The two key processes that drive the water cycle are **evaporation** and **condensation**.

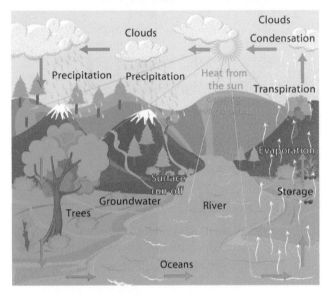

The carbon cycle

The constant recycling of carbon is called the carbon cycle.

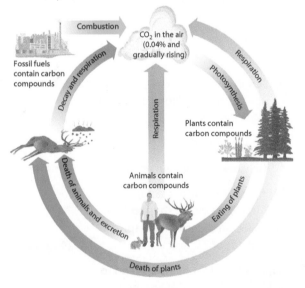

- Carbon dioxide is removed from the atmosphere by green plants for photosynthesis.
- Plants and animals respire, releasing carbon dioxide into the atmosphere.
- Animals eat plants and other animals, which incorporates carbon into their bodies. In this way, carbon is passed along food chains and webs.
- Microorganisms such as fungi and bacteria feed on dead plants and animals, causing them to decay. The microorganisms respire and release carbon dioxide gas into the air. Mineral ions are returned to the soil through decay.
- Some organisms' bodies are turned into fossil fuels over millions of years, trapping the carbon as coal, peat, oil and gas.
- When fossil fuels are burned (combustion), the carbon dioxide is returned to the atmosphere.

Controlling decomposition

Intensive use of the land removes naturally occurring nitrates, phosphates and other essential mineral nutrients for plants. Farmers can restore these nutrients by applying inorganic fertiliser or using organic means, e.g. compost or manure. Farmers and gardeners produce compost for the soil from waste plant material.

Making compost

In order for decay microorganisms to carry out decomposition, they require:

- oxygen (as the process is aerobic)
- moisture
- a warm temperature.

he presence of any of these factors will increase the rate
of decomposition.

One of your required practicals may involve investigating
the effect of one factor on the rate of decay.

Biogas digesters

A simple biogas generator

Gas trapped beneath
metal gas holder

Gas release tap

Waste
material

Residual
'digested'
sludge

Anaerobic decay produces a mixture of methane and
carbon dioxide (called **biogas**). Biogas can be made
on a large scale using a continuous-flow method in a
digester. Organic material is added daily and the biogas
is siphoned off and stored. The remaining solid sludge
can be used as fertiliser for crops.

The biogas produced by the digesters can be:

- burned to generate electricity
- burned to produce hot water for steam central
  heating systems
- used as a fuel for buses.

SUMMARY

- Materials within ecosystems are constantly
  being recycled and used to provide the
  substances that make up future organisms.
- Two of these substances are water and
  carbon.
- Biogas digestors produce a gas that can be
  used as a fuel.

QUESTIONS

QUICK TEST

1. Which two processes drive the water cycle?

2. What is biogas?

3. What three factors are needed for
   microorganisms to carry out decomposition?

EXAM PRACTICE

1. Carbon is recycled in the environment in a
   process called the carbon cycle.

   The main processes of the carbon cycle are
   shown below:

Atmosphere

5

1

3

4

Animals

2

4

a) Name the process that occurs at
   stage 1 in the diagram.   **[1 mark]**

b) If organic material is allowed to
   decompose anaerobically it
   produces a different product.

   Which gas is formed from this
   type of decay?   **[1 mark]**

c) Name one commercial use of the
   gas produced in this way.   **[1 mark]**

Environmental change & biodiversity

Environmental change

Waste management

The human population is increasing exponentially (i.e. at a rapidly increasing rate). This is because birth rates exceed death rates by a large margin.

So the use of finite resources like fossil fuels and minerals is accelerating. In addition, waste production is going up:

- **on land**, from domestic waste in landfill, toxic chemical waste, pesticides and herbicides
- **in water**, from sewage fertiliser and toxic chemicals
- **in the air**, from smoke, carbon dioxide and sulfur dioxide.

Acid rain

When coal or oils are burned, sulfur dioxide is produced. Sulfur dioxide and nitrogen dioxide dissolve in water to produce acid rain.

Acid rain can:

- damage trees, stonework and metals
- make rivers and lakes acidic, which means some organisms can no longer survive.

The acids can be carried a long way away from the factories where they are produced. Acid rain falling in one country could be the result of fossil fuels being burned in another country.

The greenhouse effect and global warming

The diagram explains how global warming can lead to climate change. This in turn leads to lower biodiversity.

Small amount of infra-red radiation transmitted into space

Infra-red radiation re-radiated into atmosphere

Carbon dioxide and methane trap the infra-red radiation (heat) as the longer wavelengths are not transmitted as easily.

UV rays from the sun reach Earth and are absorbed

The consequences of global warming are:

- a rise in sea levels leading to flooding in low-lying areas and loss of habitat
- the migration of species and changes in their distribution due to more extreme temperature and rainfall patterns; some organisms won't survive being displaced into new habitats, or newly migrated species may outcompete native species. The overall effect is a loss of biodiversity.

WS You may be asked to evaluate methods used to address problems caused by human impact on the environment.

For example, here are some figures relating to quotas and numbers of haddock in the North Sea in two successive years.

	2009	2010
Haddock quota (tonnes)	27 507	23 381
Estimated population (thousands)	102	101

What conclusions could you draw from this data? What additional information would you need to give a more accurate picture?

Biodiversity

Biodiversity is a measure of the number and variety of species within an ecosystem. A healthy ecosystem:

- has a large biodiversity
- has a large degree of interdependence between species
- is stable.

Species depend on each other for food, shelter and keeping the external physical environment maintained. Humans have had a negative impact on biodiversity due to:

- pollution killing plants and animals
- degrading the environment through deforestation and removing resources such as minerals and fossil fuels
- over-exploiting habitats and organisms.

Only recently have humans made efforts to reduce their impact on the environment. It is recognised that maintaining biodiversity is important to ensure the continued survival of the human race.

Impact of land use

As humans increase their economic activity, they use land that would otherwise be inhabited by living organisms. Examples of habitat destruction include:

- farming
- quarrying
- dumping waste in landfill.

Peat bogs

Peat bogs are important habitats. They support a wide variety of organisms and act as carbon sinks.

If peat is burned it releases carbon dioxide into the atmosphere and contributes to global warming. Removing peat for use as compost in gardens takes away the habitat for specialised animals and plants that aren't found in other habitats.

Deforestation

Deforestation is a particular problem in tropical regions. Tropical rainforests are removed to:

- **release land for cattle and rice fields** – these are needed to feed the world's growing population and for increasingly Western-style diets

- **grow crops for biofuel** – the crops are converted to **ethanol-based** fuels for use in petrol and diesel engines. Some specialised engines can run off pure ethanol.

The consequences of deforestation are:

● There are fewer plants, particularly trees, to absorb carbon dioxide. This leads to increased carbon dioxide in the atmosphere and accelerated global warming.
● Combustion and decay of the wood from deforestation releases more carbon dioxide into the atmosphere.
● There is reduced biodiversity as animals lose their habitats and food sources.

Maintaining biodiversity

To prevent further losses in biodiversity and to improve the balance of ecosystems, scientists, the government and environmental organisations can take action.

- Scientists establish **breeding programmes** for **endangered species**. These may be captive methods where animals are enclosed, or protection schemes that allow rare species to breed without being poached or killed illegally.

- The government sets **limits** on **deforestation** and **greenhouse gas emissions**.

Environmental organisations:

- **protect and regenerate** shrinking habitats such as mangrove swamps, heathlands and coral reefs

- **conserve and replant** hedgerows around the margins of fields used for crop growth

- introduce and encourage **recycling** initiatives that reduce the volume of landfill.

SUMMARY

- **The human population is increasing exponentially. Waste production is increasing.**
- **Burning of fossil fuels contributes to acid rain and global warming.**
- **Biodiversity is the number and variety of species in an ecosystem. Maintaining biodiversity is vital.**

QUESTIONS

QUICK TEST

1. Why are tropical rainforests removed?

2. Why is high biodiversity seen as a good thing?

EXAM PRACTICE

1. In Ireland, four species of bumble bee are now endangered. Scientists are worried that numbers may become so low that they are inadequate to provide pollination to certain plants.

    a) Explain how the disappearance of these four species of bumble bee might affect biodiversity as a whole. **[3 marks]**

    b) There are many threats to bumble bees, such as the possible effect of pesticides like *neonicotinoids*; loss of suitable habitat, disappearance of wildflowers and new diseases.

    Suggest two conservation measures that might help increase numbers of bumble bees in the wild. **[2 marks]**

Energy and biomass in ecosystems

Trophic levels

Communities of organisms are organised in an ecosystem according to their feeding habits.

Food chains show:

- the organisms that consume other organisms
- the transfer of energy and materials from organism to organism.

Energy from the Sun enters most food chains when green plants absorb sunlight to photosynthesise. Photosynthetic and chemosynthetic organisms are the producers of biomass for the Earth. Feeding passes this energy and biomass from one organism to the next along the food chain.

A food chain

Green plant:
producer

Rabbit:
primary consumer

Stoat:
secondary consumer

Fox:
tertiary consumer

The arrow shows the flow of energy and biomass along the food chain.

- All food chains start with a producer.
- The rabbit is a herbivore (plant eater), also known as the primary consumer.
- The stoat is a carnivore (meat eater), also known as the secondary consumer.
- The fox is the top carnivore in this food chain, the tertiary consumer.

Each consumer or producer occupies a **trophic level** (feeding level).

- Level 1 are producers.
- Level 2 are primary consumers.
- Level 3 are secondary consumers.
- Level 4 are tertiary consumers.

Excretory products and uneaten parts of organisms can be the starting points for other food chains, especially those involving decomposers.

Decomposers

Pyramids of biomass

Pyramids of biomass deal with the dry mass of living material in the chain. The width of each trophic level can be calculated by multiplying the number of organisms by their dry mass.

| Hawk |
| Thrushes |
| Slugs |
| Lettuces |

Efficiency of energy transfer

If you know how much energy is stored in the living organisms at each level of a food chain, the efficiency of energy transfer can be calculated.

To do this, divide the amount of energy used usefully (e.g. for growth) by the total amount of energy taken in.

$$\text{energy efficiency (\%)} = \frac{\text{energy used usefully}}{\text{total energy taken in}} \times 100$$

Example

A sheep eats 100 kJ of energy in the form of grass but only 9 kJ becomes new body tissue; the rest is lost as faeces, urine or heat. Calculate the efficiency of energy transfer in the sheep.

$$\text{energy efficiency} = \frac{9}{100} \times 100$$
$$= 9\%$$

imilar efficiency calculations can be
erformed for biomass.

WS During your course you will be expected to use formulae to answer questions and to re-arrange the subject of an equation. Calculating energy efficiency is one example of when you may need to do this.

For example, you might be given the energy efficiency for a particular trophic level and asked to find out the total energy taken in. In this case:

$$\text{energy efficiency (\%)} = \frac{\text{energy used usefully}}{\text{total energy taken in}} \times 100$$

would be re-arranged to form:

$$\text{total energy taken in} = \frac{\text{energy used usefully}}{\text{energy efficiency}} \times 100$$

Energy

Biomass

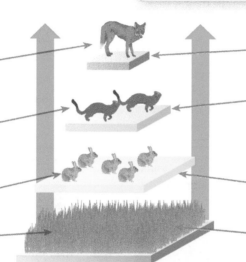

The fox gets the last tiny bit of energy left after all the others have had a share. This explains why food chains rarely have fourth degree or fifth degree consumers – they would not get enough energy to survive.

The stoats run around, mate, excrete, keep warm, etc. They pass on about a tenth of all the energy they get from the rabbits.

The rabbits run around, mate, excrete, keep warm, etc. They pass on about a tenth of all the energy they get from the grass.

The Sun is the energy source for all organisms. However, only about 10% of the Sun's energy is captured in photosynthesis.

The fox gets the remaining biomass.

The stoats lose quite a lot of biomass in faeces (egestion), urine (excretion), carbon dioxide and water (respiration).

The rabbits lose quite a lot of biomass in faeces (egestion), urine (excretion), carbon dioxide and water (respiration).

A lot of the biomass remains in the ground as the root system.

SUMMARY

- Food chains show the transfer of energy and materials from organism to organism.
- Pyramids of biomass show the overall dry mass of organisms in different trophic levels.
- Efficiency of energy transfer can be calculated by working out how much energy is used usefully.

QUESTIONS

QUICK TEST

1. What is a trophic level?
2. List three ways in which energy is lost in a food chain.

QUESTIONS

EXAM PRACTICE

1. a) The following organisms are found in a rock pool on a beach:

   Seaweed (producer/plant)

   Sea snails (herbivores)

   Starfish (carnivore)

   Using this information, sketch a likely pyramid of biomass for these organisms. **[2 marks]**

   b) Explain why the pyramid has the shape that it does. **[3 marks]**

Farming and food security

Farming techniques

Intensive farming is characterised by:

- high yields
- use of mechanised planting and harvesting
- extensive use of inorganic fertilisers, pesticides and herbicides.

In order to maximise profits, food production efficiency is the priority. This is achieved by:

- restricting energy transfer from animals to the environment, e.g. keeping animals under cover
- limiting animals' movement
- controlling the temperature of the animals' surroundings.

For example, battery chickens and calves are placed in cages or pens. Fish can be grown in cages and fed high-protein diets.

Many people object to these practices because they think animal welfare has not been given a high enough priority.

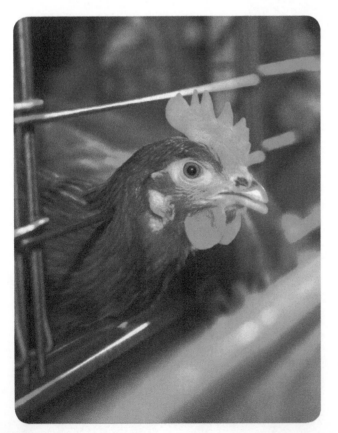

Eutrophication

Overusing fertilisers in intensive farming can lead to eutrophication.

1   Fertilisers or sewage can run into the water and pollute it. As a result, there are a lot of nitrates and phosphates, which leads to rapid growth of algae.

Algae

2   The algal blooms reproduce quickly, then die and rot. They also block off sunlight, which causes underwater plants to die and rot.

Rotting plants

3   The number of aerobic bacteria increase. As they feed on the dead organisms they use up oxygen. This causes larger organisms and plants to die because they are unable to respire.

Food security

Food security is a global issue. It recognises the importance of allowing the world's growing population to have access to an adequate diet. Greater industrialisation and higher standards of living are leading to changing diets in countries such as China. So-called 'Western diets' have a higher red meat content. This increases the demand for more land devoted to raising livestock.

Sustainable methods of food production are constantly sought, but many factors need to be taken into account when trying to achieve food security.

Many developed countries have a demand for scarce food resources, meaning that some foods are transported many thousands of miles from their point of origin.

Due to the massive use of pesticides, new pests develop that are resistant to existing chemicals. Pathogenic microorganisms also evolve to become resistant to methods of control such as antibiotics.

Climate change means that some countries are deprived of rainfall, resulting in a failure of harvests. This leads to widespread famine and armed conflicts as populations migrate to find work and food.

Cost of farming methods continues to rise.

In the future, agricultural solutions to the increasing problem of food security might include:

the increased use of hydroponics – a way of growing plants without soil under very controlled conditions

biological control – where natural invertebrate predators are used to control pests

gene technology – including pest-resistant GM crops

increased use of fertilisers and pesticides.

Sustainable fishing

Fish stocks are declining across the globe. The problem is so big that unless methods are used to halt the decline, some species (such as the northwest Atlantic cod) might disappear.

Methods to make fishing sustainable include:

governments imposing **quotas** that limit the weight of fish that can be taken from the oceans on a yearly basis – this practice is not always followed and illegal fishing is difficult to combat

increasing the mesh size of nets to allow smaller fish to escape and reach adulthood, so that they can breed.

Biotechnology

Biotechnology, particularly the use of genetic engineering, has allowed new food substances to be produced relatively cheaply. Microorganisms can be grown in large industrial vats.

For instance, the fungus, Fusarium, can be used to produce **mycoprotein**, which is protein-rich and suitable for vegetarians. The fungus is grown on glucose syrup in aerobic conditions in a matter of days. Once grown, the biomass is harvested and purified.

GM crops can increase yields and add nutritional value by increasing the vitamin content of crops such as golden rice.

SUMMARY

Intensive farming produces high yields but also has drawbacks, including questions around animal welfare, and the overuse of fertilisers causing eutrophication.

Sustainable methods of food production, such as sustainable fishing and biotechnology, are vital in ensuring the world's population has enough to eat.

QUESTIONS

QUICK TEST

1. What is eutrophication?

2. What are the advantages and disadvantages of keeping battery hens for food production?

EXAM PRACTICE

1. A farmer breeds cattle for beef. The animals are kept in enclosures inside a barn so that the temperature can be regulated and they cannot move around too much.

   a) In terms of food production, explain why the farmer would want to…

      i) regulate the temperature of the environment in which the cows are kept.

      ii) restrict how much the cattle can move around. **[2 marks]**

   b) Some people object to livestock being raised in this way.

      Suggest a reason for this. **[1 mark]**

Answers

Page 5

QUICK TEST

1. False. Eukaryotic cells are more complex than prokaryotic cells.
2. False – most cells do have a nucleus, but others, e.g. red blood cells, bacterial cells, do not.
3. Ribosomes

EXAM PRACTICE

1. a) Plasmids [1]
   b) They possess a cell wall. [1]
   c) Muscle cells require a lot of energy [1] mitochondria release this energy (for contraction) [1]
   d) 10/500 = 0.002 × 1000
      = 2 μm [**2 marks for correct answer. If incorrect, 1 mark for correct working.**]

Page 7

QUICK TEST

1. Ciliated epithelial cells
2. Xylem
3. Meristems

EXAM PRACTICE

1. Differentiation [1]
2. a) **One from:** Long, slender axons for carrying impulses long distances; Many dendrites for connecting with other nerve cells [1]
   b) **Two from:** Therapeutic cloning; Treating paralysis; Repairing nerve damage; Cancer research; Growing new organs for transplantation [2]
   c) **Two from:** Stem cells are sometimes obtained from human embryos; They believe it is wrong to use embryos (which might be deemed living) for these purposes; Risk of viral infections [2]

Page 9

QUICK TEST

1. Light microscopes and electron microscopes
2. Food/nutrients, warmth and moisture/water

EXAM PRACTICE

1. a) 3.14 × 4 × 4 = 50.2 [**2 marks for correct answer. If incorrect, 1 mark for correct working**]
   b) 851.1 ÷ 8 = 106.4 mm^2 [**2 marks for correct answer. If incorrect, 1 mark for correct working**]
2. a) Objective lenses [1]
   b) Mitochondria are very small [1] Resolving power of light microscope is not great enough [1]

Page 11

QUICK TEST

1. Gametes
2. Chromosomes
3. 23

EXAM PRACTICE

1. a) Meiosis [1]
   b) Four cells produced (in second meiotic division) [1] Each cell is haploid (not diploid) [1]

Page 13

QUICK TEST

1. Aerobic
2. A reaction that gives out heat
3. Lactic acid
4. A catabolic reaction

EXAM PRACTICE

1. a) Glucose → Lactic acid + energy released [1] Glucose → Carbon dioxide + Ethanol + Energy released [1]
   b) **1 mark must be one from:** Ethanol excreted/ removed from cell; But cannot be removed from surroundings; Humans sometimes extract it for the alcoholic drinks industry [1]
      Two from: Lactic acid causes oxygen debt; Increased breathing and heart rates deliver more blood to muscles; Oxygen removes lactic acid; Lactic acid oxidised [2]

Page 15

QUICK TEST

1. pH and temperature
2. Lipases
3. Amino acids

EXAM PRACTICE

1. a) **Three from:** High temperature denatures enzyme; Active site shape changed; Starch no longer fits active site; No maltose formed [3]
   b) **Three from:** Acidic pH keeps active site in correct shape; Protein substrate fits active site; Enzyme puts a strain on protein molecule/breaks it down; To products/peptides/amino acids [3]

Page 17
QUICK TEST
. To calculate rates of diffusion
. The plant cells have a higher water potential/lower solute concentration. The water moves **down** an osmotic gradient from inside the cells into the salt solution by osmosis.
XAM PRACTICE
. **Two from:** Ions enter by active transport; requiring release of energy; from respiration; via protein carriers in the cell membrane **[2]**
. **Four from:** The cell has become plasmolysed; The outside solution has lower water potential/higher solute concentration than cell contents; Water leaves cell; By osmosis; Down concentration/water potential/osmotic gradient **[4]**

Page 19
QUICK TEST
. The leaf
. Lignin
XAM PRACTICE
. a) **One from:** Xylem consist of hollow tubes while phloem have sieve end-plates; Xylem vessels do not possess companion cells. **[1]**
  b) Phloem carries sugar/sucrose **[1]**
     Aphids need sugar for diet/have well-adapted mouthparts to reach source of sugar. **[1]**

Page 21
QUICK TEST
. It would increase the (evapo)transpiration rate.
. Guard cells
. Potometer
XAM PRACTICE
. a) 1st 60 mins: Bubble moves 0.2 cm (2.3 – 2.1) **[1]**
     Therefore rate is 0.2 / 60 = 0.003 cm per min. **[1]**
     2nd 60 mins: Bubble moves 3.4 cm **[1]**
     Therefore rate is 3.4 / 60 = 0.06 cm per min. **[1]**
     [Correct answer in each case gains full 2 marks automatically.]
  b) **Three from:** Moving air increases rate of water absorption; Water vapour lost more rapidly from leaves/faster transpiration; Moving air increases diffusion gradient between leaf interior and atmosphere; Evaporation/diffusion is therefore more rapid **[3]**

Page 23
QUICK TEST
1. Veins
2. True
3. The pulmonary vein
EXAM PRACTICE
1. a) Artery has to withstand/recoil with higher pressure, elasticity allows smoother blood flow/second boost to blood when recoils **[1]**
   b) Vein has valves to prevent backflow of blood/compensate for low blood pressure **[1]**
2. a) The blood passes through the heart twice for every complete circuit of the body. **[1]**
   b) **One from:** Allows a higher pressure to be maintained around the body; Allows maximum efficiency for delivering oxygen/materials to body cells. **[1]**

Page 25
QUICK TEST
1. Red blood cells, white blood cells, plasma and platelets
2. Oxyhaemoglobin
EXAM PRACTICE
1. **Two from:** Involved in defence against infection/part of immune system; Some engulf foreign cells; Others produce antibodies **[2]**
2. **Three from:** Tar layer increases diffusion path for oxygen; Less oxygen is absorbed into the bloodstream; Less oxygen is delivered to respiring cells; Breathing rate is increased to compensate for this **[3]**

Page 27
QUICK TEST
1. Photosynthesis takes **in** carbon dioxide and water, and requires an energy input. It produces glucose and oxygen. Respiration **produces** carbon dioxide and water, and releases energy. It absorbs glucose and oxygen.
2. It is needed for cell walls.
EXAM PRACTICE
1. a) Increased light intensity increases rate of photosynthesis **[1]** More photosynthesis means more carbohydrate produced in a given time **[1]** More carbohydrate means bigger/more tomatoes **[1]** She should install more powerful lighting/artificial lighting at night **[1]**
   b) **Two from:** Rate of photosynthesis does not increase after a certain light intensity; Therefore any extra investment in lighting will not result in any more yield; Money will be wasted **[2]**

Answers

QUICK TEST

1. Body mass index
2. A microorganism that causes disease

EXAM PRACTICE

1. **Two from:** Emphysema; Heart disease; Stroke; **Any other reasonable answer**. [2]
2. **a)** $64 \div 1.55^2 = 26.6$ (1 d.p.) [2] **[If incorrect, allow 1 mark for correct working]**
   b) She needs to lose some weight [1] Her BMI category is 'overweight.' [1]

Page 31
QUICK TEST

1. Cholera bacteria are found in human faeces, which contaminate water supplies.
2. **Any reasonable answer, e.g.:** athlete's foot

EXAM PRACTICE

1. **a)** Pathogens [1]
   b) Viruses cause cell damage [1]; Toxins are produced that damage tissues [1]
2. Warm temperatures are ideal for mosquito to thrive [1] Stagnant water is ideal habitat for mosquito eggs to be laid/larvae to survive. [1]

Page 33
QUICK TEST

1. Phagocyte engulfs a pathogen then digests it using the cell's enzymes.
2. An antigen is a molecular marker on a pathogen cell membrane that acts as a recognition point for antibodies.
3. Lysozyme

EXAM PRACTICE

1. **a)** Antibodies [1]
   b) i) Lymphocyte [1]
       ii) Antigen is identified by the immune system [1] Memory cells are activated [1] A large amount of antibodies is produced rapidly to combat the fresh invasion of pathogens. [1]

Page 35
QUICK TEST

1. An antibiotic is a drug that kills bacteria.
2. Active and passive
3. Do not prescribe antibiotics for viral/non-serious infections; Ensure that the full course of antibiotics is completed.

EXAM PRACTICE

1. Dead/inactive form of virus is injected into person [1] this triggers production of antibodies [1] and formation of memory cells [1] If actual infection occurs, antibodies produced rapidly to deal with the pathogen [1]
2. **a)** Inhibit cell processes in bacteria [1]
   b) Two from: Antibiotics do not work against viruses/only work against bacteria; Flu caused by a virus; Over prescription risks antibiotic resistance [2]
3. MRSA is resistant to most modern antibiotics [1] caused by doctors over-prescribing antibiotics [1] new antibiotics need to be developed, which is a slow process/takes many years. [1]

Page 37
QUICK TEST

1. A double blind trial is when both volunteers and doctors don't know whether the volunteers have been given the new drug or a placebo.
2. True
3. **Two from:** Cancer treatment; Pregnancy testing kits; Measuring hormone levels; Research

EXAM PRACTICE

1. **a) Two from:** To ensure that the drug is actually effective/more effective than placebo; To work out the most effective dose/method of application; To comply with legislation; To see if they are safe / make sure that there are no dangerous side effects [2]
   b) One from: Computer modelling; Experimenting on cell cultures [1]
   c) Antibodies sensitised to the hormone HCG are bound to a test strip [1] Strip changes colour when exposed to urine from a pregnant woman [1]

Page 39
QUICK TEST
1. **Three from:** Stunted growth; Leaf spots; Areas of decay; Growths/tumours; Malformed stems and leaves; Discolouration; Presence of pests
2. Fungal disease
3. To form chlorophyll, for photosynthesis

EXAM PRACTICE
1. **Three from:** Make up 'acid rain' solution (e.g. a dilute sulfuric or nitric acid solution, or a solution with a pH in the range 2-6); Minimum of 2 groups of infected roses – spray one group with acid solution; Control variable e.g. same place in garden; Measurement/assessment technique e.g. number of bushes with black spot infection [3]
2. Mechanical: **One from:** Thorns/hairs; Leaves that droop/curl on contact; Mimicry [1]
   Physical: **One from:** Tough, waxy leaves; Cellulose cell walls; Layers of dead cells on stems [1]

Page 41
QUICK TEST
1. 37°C
2. **Two from:** Osmoregulation/water balance; Balancing blood sugar levels; Maintaining a constant body temperature; Controlling metabolic rate

EXAM PRACTICE
1. Pancreas – Produces insulin
   Skin receptor – Detects pressure
   Pituitary gland – Releases TSH
   Hypothalamus – Releases TRH
   [3 marks for 4 correct, 2 marks for 3 correct, 1 mark for 2 correct]

Page 43
QUICK TEST
1. Synapses
2. They ensure a rapid response to a threatening/harmful stimulus, e.g. picking up a hot plate. As a result, they reduce harm to the human body.

EXAM PRACTICE
1. a) X is the stimulus [1]
   b) It ensures a rapid response to a threatening/harmful stimulus [1] It is an unconscious reaction [1]
   c) **Four from:** Pain/pressure receptor in skin stimulated; Sensory neurone sends impulse; To spine; Intermediate neurone relays impulse to motor neurone; Motor neurone sends impulse to arm muscle; Arm muscle contracts [4]

Page 45
QUICK TEST
1. Central nervous system
2. Vasodilation is when blood vessels widen causing blood to flow closer to the skin's surface.

EXAM PRACTICE
1. **Six from:** Thermoregulatory centre/hypothalamus detects a rise in core temperature; Nervous signals/impulses are sent from this centre; Sweat glands release more sweat; Sweat evaporates; Endothermic change; Absorbs heat from skin; Vasodilation occurs/arterioles or blood vessels near surface of skin increase in diameter; More blood flows near surface of skin; More heat lost by radiation; When core temperature falls negative feedback reverses these changes [6]

Page 47
QUICK TEST
1. The retina
2. An eyeball that is too long, or weak suspensory ligaments

EXAM PRACTICE
1. a) Pupil decreases in diameter/size [1] due to contraction of the iris muscle [1] to prevent light damaging the retina [1]
   b) **Three from:** Ciliary body/muscle relaxes; Stretching lens more thinly/lens adopts less convex shape; Light rays refracted less; In order to converge/focus at a sharp point on the retina [3]

Page 49
QUICK TEST
1. Thyroxine
2. Pancreas
3. Reduced ability of cells to absorb insulin and therefore high levels of blood glucose. This leads to tiredness, frequent urination, poor circulation, eye problems, etc.

EXAM PRACTICE
1. a) **Two from:** Causes all body cells to absorb glucose; Causes liver to convert glucose to glycogen; Blood glucose level is reduced [2]
   b) Insulin injections and monitoring of diet [1]
   c) The body's cells often no longer respond to insulin [1] It can be managed by adjusting carbohydrate intake [1]

Answers

Page 51

QUICK TEST

1. **Three from:** Glucose; Amino acids; Fatty acids; Glycerol; Some water
2. The pituitary gland

EXAM PRACTICE

1. a) Respiration [1]
   b) The lungs [1]
   c) Amino acids [1]

Page 53

QUICK TEST

1. FSH, LH, progesterone, oestrogen
2. **One from:** Production of sperm in testes; Development of muscles and penis; Deepening of the voice; Growth of pubic, facial and body hair

EXAM PRACTICE

1. a) Oestrogen – promotes repair of uterus wall and stimulates egg release [1]
   Progesterone – maintains lining of uterus [1]
   b) Release of egg/ovulation [1]
2. Breast development [1] Sperm production [1] Menstrual cycle [1]

Page 55

QUICK TEST

1. Spermicidal agent. It is cheap.
2. They provide a barrier/prevent transferral of the virus during sexual intercourse.

EXAM PRACTICE

1. Advantage – very effective against STIs [1]
   Disadvantage – **One from:** Can only be used once; May interrupt sexual activity; Can break; May be allergic to latex [1]
2. Woman is given FSH and LH to stimulate the production of several eggs [1] Sperm and eggs are then introduced together outside the body in a petri dish [1] Successfully growing embryos can then be transplanted into the woman's uterus [1]

Page 57

QUICK TEST

1. Positive phototropism
2. Auxin
3. Gibberellins

EXAM PRACTICE

1. Plant hormones [1]
2. Light causes auxin to build up on lower side of shoot/ side of shoot away from light [1] This causes greater cell elongation on this side of the shoot [1]

Page 59

QUICK TEST

1. Fertilisation
2. Via runners
3. By spores

EXAM PRACTICE

1. Advantage – **One from:** Produces variation; Survival advantage when the environment is changing; Can be used by humans for selective breeding purposes [1]
   Disadvantage: **One from:** Slow process; Disadvantage in stable environments; More resources required; Results of selective breeding are unpredictable [1]
2. a) Not as dependent on one species/if one species is low in numbers or becomes extinct there is another organism available to act as a 'host.' [1]
   b) **One from:** Fungi; Mosses; Ferns; Protista [1]

Page 61

QUICK TEST

1. It has allowed the production of linkage maps that can be used for tracking inherited traits from generation to generation. This has led to targeted treatments for these conditions.
2. A, T, C and G

EXAM PRACTICE

1. By collecting and analysing DNA samples from different groups of peoples [1] human migration patterns can be deduced [1]
2. Backbone consisting of a sugar molecule [1] attached to a phosphate molecule [1] nitrogenous base attached [1]

Page 63

QUICK TEST

1. Dominant and recessive
2. messenger RNA (involved in transcription) and transfer RNA (involved in translation)

EXAM PRACTICE

1. A recessive allele controls the development of a characteristic only if a dominant allele is not present. [1] So, the recessive allele must be present on both chromosomes in a pair. [1]
2. Messenger RNA – Constructed in the nucleus [1] Forms a complementary strand to DNA [1] Transfer RNA – Found in the cytoplasm (or temporarily in the ribosome) [1] Carry amino acids to add to a forming protein [1]

QUICK TEST

. 0%

EXAM PRACTICE

. Parents: Ff   x   FF   [1]
  Gametes: F   f   x   F   [1]

Offspring   FF   Ff   [1]
Phenotypes Unaffected   Carrier   [1]
           50   :   50   [1]

No children will suffer from cystic fibrosis;
50% chance of a child being a carrier [1]

Page 67
QUICK TEST

. **Two from:** The fossil record; Comparative anatomy (pentadactyl limb); Looking at changes in species during modern times; Studying embryos and their similarities; Comparing genomes of different organisms

. **Two from:** Catastrophic events, e.g. volcanic eruption; Changes to the environment; New predators; New diseases; New competitors

EXAM PRACTICE

. Variation – horse-like ancestor adapted to environment/had different characteristics, named examples of different characteristics, [1] e.g. some horse-like mammals had more flipper-like limbs (as whales have flippers), mutation in genes allowed some individuals to develop these advantageous characteristics [1]

Idea of competition for limited resources; examples of different types of competition e.g. obtaining food in increasingly water-logged environment [1]

Idea of survival of the fittest – named examples of different adaptations, e.g. some horse-like mammals had more flipper-like limbs that allowed them to swim well in water [1]

Idea of successful characteristics being inherited – genes passed on to next generation, extinction – variety which is least successful does not survive and becomes extinct [1]

Page 69
QUICK TEST

1. The biological changes seen in species over millions of years
2. Ardi
3. **Any two examples**, e.g. Peppered moths; Antibiotic resistance in bacteria

EXAM PRACTICE

1. **Two from:** Changes in DNA/genes; Mutation; Production of new proteins; Proteins determine structural or behavioural adaptations [2]

Page 71
QUICK TEST

1. In Darwin's theory, change in the inherited material occurs first. This leads to change in the phenotype/external appearance. Lamarck's theory stated that the physical change occurs first.
2. Cloning allows mass production of individuals with beneficial traits in a short space of time.

EXAM PRACTICE

1. – Scrape off small pieces of tissue into vessels containing nutrient and hormones [1]
   – Remove plantlets/clones [1]
   – Repeat process [1]
2. It is possible to mass produce organisms with specific characteristics beneficial to humans [1] Stated example: e.g. High meat quality in cattle/pigs/sheep; Greater milk yields in dairy cattle; High quality wool in sheep, etc. [maximum 2]

Answers

QUICK TEST
1. Genetic engineering is more precise and it takes less time to see results.
2. A plasmid is a ring of DNA found in bacteria.

EXAM PRACTICE
1. Advantage – **One from:** It results in an organism with the 'right' characteristics for a particular function; In farming and horticulture, it is a more efficient and economically viable process than natural selection [1]
Disadvantage – **One from:** Intensive selective breeding reduces the gene pool – the range of alleles in the population decreases so there is less variation; Lower variation reduces a species' ability to respond to environmental change; It can lead to an accumulation of harmful recessive characteristics (in-breeding) [1]
2. Stage 2: Gene removed using restriction enzyme [1]
Stage 3: Bacterial plasmid 'cut open' using restriction enzyme [1]
Stage 4: Ligase enzyme used to insert human gene into bacterial plasmid [1]
Stage 5: Bacterial cells containing plasmids rapidly reproduce [1]
Stage 6: Insulin purified from fermenter culture and produced in commercial quantities [1]

Page 75
QUICK TEST
1. Genus and species
2. Carl Woese

EXAM PRACTICE
1. Phylum [1] Order [1] Genus [1]
2. Canis familiaris [1]

Page 77
QUICK TEST
1. An ecosystem includes the habitat, its communities o animals and plants, together with the physical factors that influence them.
2. Food, mates, territory
3. **Any two suitable examples**, e.g. bacteria living in deep sea vents and icefish (exist in waters less than 0°C in temperature).

EXAM PRACTICE
1. i) Allow it to grip/stick to trees [1]
   ii) Camouflage so it won't be seen [1]
   iii) To catch insects/prey [1]
2. Extremophiles are organisms that can survive in extreme environmental conditions [1]
   One from: Bacteria living in deep sea vents have optimum temperatures for enzymes that are much higher than 37°C; Icefish have antifreeze chemicals in their bodies which lower the freezing point of body fluids; Some organisms resist high salt concentrations or pressure [1]

Page 79
QUICK TEST
1. It would tell you how numbers of different species vary along the line of the transect.
2. Biotic factors are living; abiotic factors are non-living

EXAM PRACTICE
1. a) To ensure they obtained representative samples [1]
   b) Working – numbers arranged in order of magnitude: 0,0,1,2,2,3,4,5,5,6
   The middle 2 numbers are 2 and 3, therefore the median is 2.5 **[3 marks for correct answer. If incorrect, 1 mark for each correct working]**
   c) Area of the field is 60 x 90 = 5400 m^2
   Therefore, number of daisies = 5400 × 2.3 = 12 42(**[2 marks for correct answer. If incorrect, 1 mark for correct working]**

Page 81

QUICK TEST
1. An organism used as an indicator of pollution
2. An animal that kills and eats other animals

EXAM PRACTICE
1. a) Yes [0 marks] There are no species present that you might find in very polluted water e.g. rat tailed maggot, bloodworm [1] There are many 'clean-water' indicator species e.g. stonefly nymphs [1]
   b) Direct chemical tests e.g. pH of the water, oxygen concentrations [1]
2. As the rabbit population decreases, there is less food available for the stoats [1] so the stoat population decreases [1] fewer rabbits are eaten, so the rabbit population increases [1]

Page 83

QUICK TEST
1. Condensation and evaporation
2. A mixture of methane and carbon dioxide
3. Oxygen, warmth, moisture

EXAM PRACTICE
1. a) Photosynthesis [1]
   b) Methane [1]
   c) **One from:** Electricity generation; Generation of steam for central heating systems; Used as bus fuel [1]

Page 85

QUICK TEST
1. **One from:** To provide land for cattle and rice fields; To grow crops for biofuel
2. With more species, there are more relationships between different organisms and therefore more resistance to disruption from outside influences.

EXAM PRACTICE
1. a) **Three from:** Certain plant species/crops become scarce; Due to lack of pollinators; Other species which depend on these plants will be endangered/reduced in number; Biodiversity falls [3]
   b) **Two from:** Reduce usage of pesticides such as *neonicotinoids*; Plant more species which attract bumble bees (flowers, etc.); Farm bees in hives and release into the wild; Disease prevention measures to act against foul-brood disease/isolate colonies/remove diseased colonies [2]

Page 87

QUICK TEST
1. The level that an organism feeds at, e.g. producer, primary consumer level. It refers to layers in a pyramid of biomass.
2. **Three from:** Respiration; Movement; Reproduction; Maintaining constant body temperature; Urine; Faeces; Not all parts of the animal/plant are eaten.

EXAM PRACTICE
1. a)

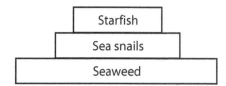

   [1 mark for correct sequence of organisms, 1 mark for correct shape of pyramid]
   b) Energy lost between trophic levels [1] through respiration, movement of organism, body heat, undigested food [1] so less energy available for next trophic level [1]

Page 89

QUICK TEST
1. A process where nitrates and phosphates enrich waterways, causing massive growth of algae and loss of oxygen.
2. **Advantages:** cheap cost of production; rapid growth of stock **Disadvantage:** reduced animal welfare

EXAM PRACTICE
1. a) i)   So energy is not wasted by the cows generating heat to keep warm [1]
      ii) So energy is not lost through movement of cows [1]
   b) They believe it is cruel because the cows cannot behave in a natural way. [1]

Glossary

Active ingredient – Chemical in a drug that has a therapeutic effect (other chemicals in the drug simply enhance flavour or act as bulking agents)

Active site – The place on an enzyme molecule into which a substrate molecule fits

Alleles – Alternative forms of a gene on a homologous pair of chromosomes

HT Ammonia – Nitrogenous waste product

Anatomy – The study of structures within the bodies of organisms

Anthropologists – Scientists who study the human race and its evolution

Antibodies – Proteins produced by white blood cells (particularly lymphocytes). They lock on to antigens and neutralise them

Antigen – Molecular marker on a pathogen cell membrane that acts as a recognition point for antibodies

Aphid – A type of sap-sucking insect

Binary fission – Process where bacterial cells divide into two

Binomial system – 'Two name' system of naming species using Latin

Biomass – Mass of organisms calculated by multiplying their individual mass by the number that exist

Biosphere – Area on the Earth's crust that is inhabited by living things

Carbon sinks – Resources that lock up carbon in their structure rather than allowing them to form carbon dioxide, e.g. peat bogs, oceans, limestone deposits

Catalyst – A substance that controls the rate of a chemical reaction without being chemically changed itself

Cellulose – Large carbohydrate molecule found in all plants; an essential constituent of cell walls

Chlorophyll – A molecule that gives plants their green colour and absorbs light energy

Chlorosis – Where leaves lose their colour as a result of mineral deficiency

Cilia – Microscopic hairs found on the surface of epithelial cells; they 'waft' from side to side in a rhythmic manner

Clone – A genetically identical cell, tissue, organ or organism

Colony – A large number of microorganisms (of one type) growing in a location, e.g. a circular colony on the surface of agar

Common ancestor – Organism that gave rise to two different branches of organisms in an evolutionary tree

Competition – When two individuals or populations seek to exploit a resource, e.g. food. One individual/population will eventually replace the other

HT Complementary strand – A sequence of bases that can bond with an equivalent sequence in DNA. Complementary strands can be DNA or mRNA

Compost – Fertiliser produced from the decay of organic plant material

Compress – Squash or squeeze. In geology, this is usually due to Earth movements or laying down sediments

Concave – Lens that curves inwards on one or both sides

Contraception – Literally means 'against conception'; any method that reduces the likelihood of a sperm meeting an egg

Converge – Come together at a point

Convex – Lens that curves outwards on one or both sides

Coronary artery – The blood vessel delivering blood to the heart muscle

Daughter cells – Multiple cells arising from mitosis and meiosis

HT Deamination – Process in the liver in which nitrogen is removed from an amino acid molecule

Decomposers – Microorganisms that break down dead plant and animal material by secreting enzymes into the environment. Small, soluble molecules can then be absorbed back into the microorganism by the process of diffusion

Denaturation – When a protein molecule such as an enzyme changes shape and makes it unable to function

Dermis – The layer of tissue in the skin that lies beneath the epidermis. It contains many receptors and structures involved in temperature control

HT Diagnostic – A process used to identify or recognise a disease or pathogen

Differentiation – A process by which cells become specialised, i.e. have a particular structure to carry out their function

Glossary

Diploid – A full set of chromosomes in a cell (twice the haploid number)

Diverge – Spread outwards from a point

Droplet infection – Transmission of microorganisms through the aerosol (water droplets) produced through coughing and sneezing

Ductless gland – A gland that does not secrete its chemicals through a tube. The pancreas is an exception to this rule as it contains a duct for delivering enzymes, but its hormones are released directly into the bloodstream

Emissions – Gaseous products usually connected with pollution, e.g. carbon dioxide emissions from exhausts

Endangered – Category of risk attached to rare species of plants and animals. This usually triggers efforts to preserve the species' numbers

Endocrine system – System of ductless organs that release hormones

Endothermic – A change that requires the input of energy

Energy demand – Energy required by tissues (particularly muscle) to carry out their functions

Environmental resources – Materials or factors that organisms need to survive, e.g. high oxygen concentration, living space or a particular food supply

Epidermis – Outer layer of tissue in the skin

Epithelial – A single layer of cells often found lining respiratory and digestive structures

Eutrophication – A process where nitrates and phosphates enrich waterways, causing massive growth of algae and loss of oxygen

(Evapo)transpiration – Evaporation of water from stomata in the leaf

Exothermic reaction – A reaction that gives out heat

External fertilisation – Gametes join outside the body of the female

Extinction – When there are no more members of a species left living

Fittest – The most adapted individual or species

Follicle stimulating hormone – A hormone produced by the pituitary gland that controls oestrogen production by the ovaries

Fusion – Joining together; in biology the term is used to describe fertilisation

Gamete – A sex cell, i.e. sperm or egg

Glucagon – Hormone released by the pancreas that stimulates the conversion of glycogen to glucose

Glycogen – Storage carbohydrate found in animals

Gravitropism (geotropism) – Growth response in plants against or with the force of gravity

Haemoglobin – Iron-containing molecule that binds to oxygen molecules in red blood cells

Haploid – A half set of chromosomes in a cell; haploid cells are either eggs or sperm

Haploid nucleus – Nucleus with a half chromosome set

Herbicide – A chemical applied to crops to kill weeds

Herd immunity – Vaccination of a significant portion of a population (or herd) makes it hard for a disease to spread because there are so few people left to infect. This gives protection to susceptible individuals such as the elderly or infants

Homologous chromosomes – A pair of chromosomes carrying alleles that code for the same characteristics

HT Hybridoma cells – Cells produced by the fusion of lymphocytes and a tumour cell

Hydrogen bond – A bond formed between hydrogen and oxygen atoms on different molecules close to each other

Immune system – A system of cells and other components that protect the body from pathogens and other foreign substances

Immunological memory – The system of cells and cell products whose function is to prevent and destroy microbial infection

Implantation – Process in which an embryo embeds itself in the uterine wall

Impulses – Electrical signals sent down neurones that trigger responses in the nervous system

Indicator species – Organism used as an indicator of pollution

Inhibition – The effect of one agent against another in order to slow down or stop activity, e.g. chemical reactions can be slowed down using inhibitors. Some hormones are inhibitors

Internal fertilisation – Gametes join inside the body of the female

Intrauterine – Inside the uterus

Kinetic energy – Energy possessed by moving objects, e.g. reactant molecules such as enzyme and substrate molecules

Lignin – Strengthening, waterproof material found in walls of xylem cells

Limiting factor – A variable that, if changed, will influence the rate of reaction most

Malnutrition – A diet lacking in one or more food groups

Menstruation – Loss of blood and muscle tissue from the uterus wall during the monthly cycle

Meristems – Growth regions in plants where stem cells are produced, e.g. apical bud

Mucus – Thick fluid produced in the lining of respiratory and digestive passages

Neuroscience – Study of the nervous system

Non-communicable – Disease or condition that cannot be spread from person to person via pathogen transfer

Oral contraceptive – Hormonal contraceptive taken in tablet form

Organelle – A membrane-bound structure within a cell that carries out a particular function

Organic – Material obtained from living things; either their dead bodies or their waste. Organic can also refer to a type of farming that avoids overuse of intensive practices such as pesticides and inorganic fertilisers

Oxygen debt – The oxygen needed to remove lactic acid after exercise

Partially permeable membrane – A membrane with microscopic holes that allows small particles through (e.g. water) but not large ones (e.g. sugar)

Partially permeable membrane – Artificial or organic layer that only allows small molecules through

Pathogen – A harmful microorganism

Pesticide – Chemical sprayed on crops to kill invertebrate pests

Pharmaceutical drug – Chemicals that are developed artificially and taken by a patient to relieve symptoms of a disease or treat a condition

Photosynthetic organism – Able to absorb light energy and manufacture carbohydrate from carbon dioxide and water

Phototropism – Growth response in plants, towards or away from the direction of incoming light

Placebo – A substitute for the medication that does not contain the active ingredient

Plaque – Fatty deposits that can build up in arteries

(HT) Plasmid – A ring of DNA found in bacteria

Polymer – A long chain molecule made up of individual units called monomers

(HT) Polypeptide – A sequence of peptides, which in turn are smaller chains of amino acids

Receptor molecule – Protein on the outer membrane of a cell that binds to a specific molecule, such as transmitter substance

Regeneration – Rebuilding or regrowth of a habitat, e.g. flooding peatland to encourage regrowth of mosses and other plants

Resolution – The smallest distance between two points on a specimen that can still be told apart

Sample – A small area or population that is measured to give an indication of numbers within a larger area or population

Selective re-absorption – Occurs in the kidney tubules and allows useful substances to pass from the kidneys back into the blood

Sensitisation – Cells in the immune system are able to 'recognise' antigens or foreign cells and respond by attacking them or producing antibodies

Spinal cord – Nervous tissue running down the centre of the vertebral column; millions of nerves branch out from it

Stem cells – Undifferentiated cells that can become any type of cell

Stimuli – Changes in the internal or external environment that affect receptors

Substrate – The molecule acted on by an enzyme

Surface area to volume ratio – A number calculated by dividing the total surface area of an object by its volume. When the ratio is high, the efficiency of diffusion and other processes is greater

Glossary

Surrogate mother – An adult female animal that has had an embryo implanted into her uterus. The surrogate gives birth to the offspring but has no genetic relationship to it

Sustainability – Carrying out human activity, e.g. farming, fishing and extraction of resources from the ground, so that damage to the environment is minimised or removed

Symptoms – Physical or mental features that indicate a condition or disease, e.g. rash, high temperature, vomiting

Therapeutic cloning – A process where embryos are grown to produce cells with the same genes as a particular patient

HT **Transcription** – Process by which a complementary strand of mRNA is made in the nucleus

HT **Translation** – Process occurring in ribosomes where polypeptides are produced from the code in mRNA

Translocation – Process in which sugars move through the phloem

Transpiration – Flow of water through the plant ending in evaporation from leaves

Turgidity – Where plant cells fill with water and swell as a result of osmosis

Urea – Nitrogenous waste product

Vectors – Small organisms (such as mosquitoes or ticks) that pass on pathogens between people or places

Ventilation – Process of drawing air into and out of the lungs. It involves the ribs, intercostal muscles and diaphragm

Vertebrate – Animal with a backbone

Water potential/diffusion gradient – A higher concentration of particle numbers in one area than another; in living systems, these areas are often separated by a membrane or cell wall

Yield – The weight of living material harvested in farming and fishing